Introduction to Digital Signal Processing

Essential Electronics Series

Introduction to Digital Signal Processing

Bob Meddins
School of Information Systems
University of East Anglia, UK

OXFORD   AUCKLAND   BOSTON   JOHANNESBURG   MELBOURNE   NEW DELHI

Newnes an imprint of Butterworth-Heinemann
Linacre House, Jordan Hill, Oxford OX2 8DP
225 Wildwood Avenue, Woburn, MA 01801-2041
A division of Reed Educational and Professional Publishing Ltd

 A member of the Reed Elsevier plc group

First published 2000
Transferred to digital printing 2004
© 2000 Bob Meddins

1007238369

Whilst the advice and information in this book are believed to be true and
accurate at the date of going to press, neither the author nor the publisher
can accept any legal responsibility or liability for any errors or omissions
that may be made.

British Library Cataloguing in Publication Data
A catalogue record for his book is available from the British Library

ISBN 0 7506 5048 6

Typeset in 10.5/13.5 New Times Roman by Replika Press Pvt Ltd,
100% EOU, Delhi 110 040, India

Series Preface

In recent years there have been many changes in the structure of undergraduate courses in engineering and the process is continuing. With the advent of modularization, semesterization and the move towards student-centred learning as class contact time is reduced, students and teachers alike are having to adjust to new methods of learning and teaching.

Essential Electronics is a series of textbooks intended for use by students on degree and diploma level courses in electrical and electronic engineering and related courses such as manufacturing, mechanical, civil and general engineering. Each text is complete in itself and is complementary to other books in the series.

A feature of these books is the acknowledgement of the new culture outlined above and of the fact that students entering higher education are now, through no fault of their own, less well equipped in mathematics and physics than students of ten or even five years ago. With numerous worked examples throughout, and further problems with answers at the end of each chapter, the texts are ideal for directed and independent learning.

The early books in the series cover topics normally found in the first and second year curricula and assume virtually no previous knowledge, with mathematics being kept to a minimum. Later ones are intended for study at final year level.

The authors are all highly qualified chartered engineers with wide experience in higher education and in industry.

R G Powell
Jan 1995
Nottingham Trent University

To the memory of
my father
John Reginald (Reg) Meddins
(1914–1974)
and
our son
Huw
(1977–1992)

Contents

Preface

As early as the 1950s, designers of signal processing systems were using digital computers to simulate and test their designs. It didn't take too long to realize that the digital algorithms developed to drive the simulations could be used to carry out the signal processing directly – and so the *digital signal processor* was born. With the incredible development of microprocessor chips over the last few decades, digital signal processing has become a hugely important topic. Speech synthesis and recognition, image and music enhancement, robot vision, pattern recognition, motor control, spectral analysis, anti-skid braking and global positioning are just a few of the diverse applications of digital signal processors.

Digital signal processing is a tremendously exciting and intriguing area of electronics but its essentially mathematical nature can be very off-putting to the newcomer. My goal was to be true to the title of this book, and give a genuine *introduction* to the topic. As a result, I have attempted to give good coverage of the essentials of the subject, while avoiding complicated proofs and heavy maths wherever possible. However, references are frequently made to other texts where further information can be found, if required. Each chapter contains many worked examples and self-assessment exercises (along with worked solutions) to aid understanding. The student edition of the software package, MATLAB, is used throughout, to help with both the analysis and the design of systems. Although it is not essential that you have access to this package, it would make the topic more meaningful as it will allow you to check your solutions to some of the problems and also try out your own designs relatively quickly. I have not included a tutorial on MATLAB as there are many excellent texts that are dedicated to this. A reasonable level of competence in the use of some of the basic mathematical tools of the engineer, such as partial fractions, complex numbers and Laplace transforms, is assumed.

After working through this book, you should have a clear understanding of the principles of the digital signal processor and appreciate the cleverness and flexibility of the device. You will also be able to design digital filters using a variety of techniques. By the end of the book you should have a sound basis on which to build, if you intend embarking on a more advanced or specialized course.

Bob Meddins
Norwich, 2000

Acknowledgements

Thanks are due to the lecturers who originally introduced me to the mysteries of this topic and to the numerous authors whose books I have referred to over the years. Thanks also to the many students who have made my teaching career so satisfying – I have learned much from them.

Special thanks are due to Siân Jones of Butterworth-Heinemann, who guided me through the project with such patience and good humour. I am also indebted to the team of anonymous experts who had the unenviable task of reviewing the book at its various stages. I am grateful to Simon Nicholson, a postgraduate student at the University of East Anglia, who gave good advice on particular aspects of MATLAB, and also to several staff at the MathWorks, MATLAB helpdesk, who responded so rapidly to my e-mail cries for help!

Finally, special thanks to my wife, Brenda, and daughter, Cathryn, for their unstinting encouragement and support.

1 The basics

1.1 CHAPTER PREVIEW

In this first chapter you will be introduced to the basic principles of digital signal processing (DSP). We will look at how digital signal processing differs from the more conventional analogue signal processing and also at its many advantages. Some simple digital processing systems will be described and analysed. The main aim of this chapter is to set the scene and give a feel for what digital signal processing is all about – most of the topics mentioned will be revisited, and dealt with in more detail, in other chapters.

1.2 ANALOGUE SIGNAL PROCESSING

You are probably very familiar with *analogue* signal processing. Some obvious examples of this type of processing are amplification, rectification and filtering. With all analogue processing, signals enter a system, are processed by passing them through circuits containing capacitors, resistors, inductors, op amps, transistors etc. They are then outputted from the system with a different shape or size. Figure 1.1 shows a very elementary example of an analogue signal processing system, consisting of just a resistor and a capacitor – you will probably recognize it as a simple type of lowpass filter. Analogue signal processing circuits are commonplace and have been very important system building blocks since the early days of electrical engineering.

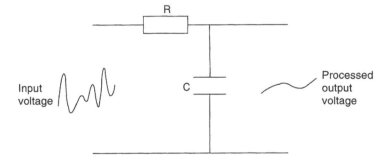

Figure 1.1

Unfortunately, as useful as they are, analogue processing systems do have major defects. An obvious one is that they have to be physically modified if the processing needs to be changed. For example, if the gain of an amplifier has to be increased, then this usually means that at least a resistor has to be changed.

What if a different cut-off frequency is required for a filter or, even worse, we want to replace a highpass filter with a lowpass filter? Once again, components must be changed. This can be very inconvenient to say the least – it's bad enough when a single system has to be adjusted but imagine the situation where a batch of several thousand is found to need modifying. How much better if changes could be achieved by altering a parameter or two in a computer program . . .

Another problem with analogue systems is that of '*repeatability*'. It is *very* unlikely that two analogue systems will have identical performances, even though they have been made in exactly the same way, with supposedly the same value components. This is mainly because of component tolerances. Analogue devices have two further disadvantages. The first is that their components age and so the device performance changes. The other is that components are also affected by temperature changes.

1.3   AN ALTERNATIVE APPROACH

So, having slightly dented the reputation of analogue processors, what's the alternative? Luckily, signal processing systems do exist which work in a completely different way and do not have these problems. A major difference is that these systems first sample, at regular intervals, the signal to be processed (Fig. 1.2). The sampled voltages are then converted to equivalent binary values, using an

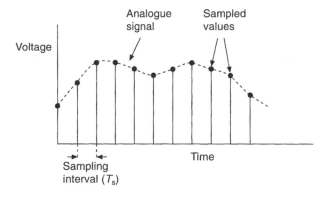

Figure 1.2

analogue-to-digital converter (Fig. 1.3). Next, these binary numbers are fed into a digital processor, containing a particular program, which will change the samples. The way in which the digital values are modified will obviously depend on the type of signal processing required – for example, do we want lowpass or highpass filtering and what cut-off frequency do we require? The transformed samples are then outputted, via a digital-to-analogue converter, to produce the reconstituted but processed analogue output signal.

Because computers can process data so quickly, the signal processing can be done almost in 'real time', i.e. the processed output samples are fed out continuously, almost in step with the corresponding input samples. Alternatively, the processed data could be stored, perhaps on a chip or CD-ROM, and then read when required.

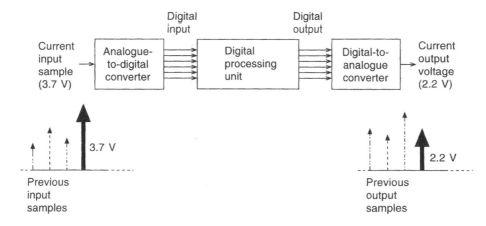

Figure 1.3

By now, you've probably guessed that this form of processing is called *digital signal processing*. Digital signal processing (DSP) does not have the drawbacks of analogue signal processing, already mentioned. For example, the type of processing required can be modified very easily – if the specification of a filter needs to be changed then new parameters can simply be keyed into the DSP system, i.e. the processing is *programmable*. The performance of a digital filter is also constant, not changing with either time or temperature. DSP systems are also inherently repeatable – if several DSP systems have been programmed to process signals in a certain way then they will all behave identically. DSP systems can also process signals in ways impossible for analogue systems.

To summarize:

- *Digital signal processing systems* are available that will do almost everything that analogue signals can do, and much more – '*versatile*'.

- They can be easily changed – '*programmable*'.

- They can be made to process signals identically – '*repeatable*'.

- They are not affected by temperature or ageing – '*physically stable*'.

1.4 THE COMPLETE DSP SYSTEM

The heart of the digital signal processing system, the analogue-to-digital converter (ADC), digital processor and the digital-to-analogue converter (DAC), is shown in Fig. 1.3. However, this sub-unit needs 'topping and tailing' in order to create the complete system. An entire, general DSP system is shown in Fig. 1.4.

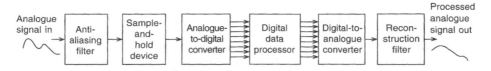

Figure 1.4

Each block will now be described briefly.

The anti-aliasing filter

If the analogue input voltage is not sampled frequently enough then this results in something of a shambles. Basically, high frequency input signals will appear as low frequency signals at the output, which will be very confusing to say the least! This phenomenon is called *aliasing*. In other words, the high frequency input signals take on another identity, or 'alias', on leaving the system.

To get a feel for the problem of aliasing, consider a sinusoidal signal, of fixed frequency, which is being sampled every 7/8 of a period, i.e. $7T/8$ (Fig. 1.5). Having only the samples as a guide, it can be seen that the sampled signal appears to have a much lower frequency than it really has.

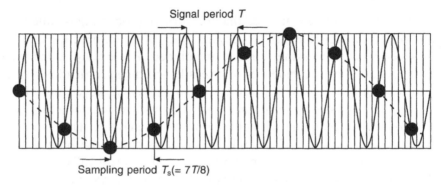

Figure 1.5

In practice, a signal will not usually have a single frequency but will consist of a very wide range of frequencies. For example, audio signals can contain frequency components in the range of about 20 Hz to 20 kHz.

To prevent aliasing, it can be shown that the signal must be sampled at least twice as fast as the highest frequency component.

This very important rule is known as the Nyquist criterion, or Shannon's sampling theorem, after two distinguished pioneers from the world of signal processing.

If this sampling rate cannot be achieved, perhaps because the components used just cannot respond this quickly, then a lowpass filter must be used on the input end of the system. This has the job of removing signal frequencies greater than $f_s/2$, where f_s is the sampling frequency. This is the role of the *anti-aliasing filter*. An anti-aliasing filter is therefore a lowpass filter with a cut-off frequency of $f_s/2$.

The important frequency of $f_s/2$ is usually called the *Nyquist frequency*.

The sample-and-hold device

An ADC should not be presented with a changing voltage to convert. The changing signal should be sampled and then this sampled voltage held while the conversion is carried out (Fig. 1.6). (In practice, the sampled value is normally held until the next sample is taken.) If the voltage is *not* kept constant during conversion then, depending on the type of converter used, the digital output might not just be a little inaccurate but could be absolute rubbish, bearing no relationship to the true value.

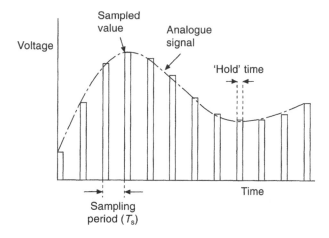

Figure 1.6

At the heart of the *sample-and-hold* device is a capacitor (Fig. 1.7). The electronic switch, S, is closed, causing the capacitor to charge to the current value of the input voltage. After a brief time interval the switch is reopened, so keeping the sampled voltage across the capacitor constant while the ADC carries out its conversion. The complete sample-and-hold device usually includes a voltage follower at both the input and the output of the basic system shown in Fig. 1.7. The characteristically low output impedance and high input impedance of the voltage followers ensure that the capacitor is charged very quickly by the input voltage and discharges very slowly through the ADC connected to its output, so maintaining the stored voltage.

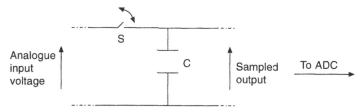

Figure 1.7

The analogue-to-digital converter

This converts the steady, sampled voltage, supplied by the sample-and-hold device, to an equivalent digital value in preparation for processing. The more output bits the converter has, the finer the resolution of the device, i.e. the smaller is the voltage change represented by the least significant output bit changing from 0 to 1 or from 1 to 0.

You are probably aware that there are many different types of ADC available. However, some of these are too slow for most DSP applications, e.g. single- and dual-slope and the basic counter-feedback versions. An ADC widely used in DSP systems is the sigma-delta converter. If you feel the need to do some extra reading in order to brush up on ADCs then some keywords to look out for are: single-slope, dual-slope, counter-feedback, successive approximation, flash, tracking and sigma-delta converters and also converter resolution. Millman and Grabel (1987) is just one of many books that give a good general treatment, while Marven and Ewers (1994) and also Proakis and Manolakis (1996) are two texts that give good coverage of the sigma-delta converter.

The processor

This *could* be a general-purpose microprocessor chip, but this is unlikely. The data processing part of a purpose-built DSP chip is designed to be able to do a limited number of fairly simple operations, in particular addition and multiplication, *but they do these exceptionally quickly*. Most of the major chip-producing companies have developed their own DSP chips, e.g. Motorola, Texas Instruments and Analog Devices, and their user manuals are obvious reference sources for further reading.

The digital-to-analogue converter

This converts the processed digital value back to an equivalent analogue voltage. Common types are the 'weighted resistor' and the 'R-2R' ladder converters, although the weighted resistor version is not a practical proposition, as it cannot be fabricated sufficiently accurately as an integrated circuit. Details of these two devices can be found in Millman and Grabel (1987), while Marven and Ewers (1994) describes the more sophisticated 'bit-stream' DAC, often used in DSP systems.

The reconstruction filter

As the anti-aliasing filter ensures that there are no frequency components greater than $f_s/2$ entering the system, then it seems reasonable that the output signal will also have no frequency components greater than $f_s/2$. However, this is not so! The output from the DAC will be 'steppy' because the DAC can only output certain voltage values. For example, an 8-bit DAC will have 256 different output voltage levels going from perhaps –5 V to +5 V. When this quantized output is analysed, frequency components of f_s, $2f_s$, $3f_s$, $4f_s$ etc. (harmonics of the sampling frequency) are found. The very action of sampling and converting introduces these harmonics of the sampling frequency into the output signal. It is these harmonics which give the output signal its steppy appearance. The *reconstruction filter* is a lowpass filter having a cut-off frequency of $f_s/2$, and is used to filter out these harmonics and so smooth the output signal.

1.5 RECAP

- Analogue signal processing systems have a variety of disadvantages, such as components needing to be changed in order to change the processor function, inaccuracies due to component ageing and temperature changes, processors built in the same way not performing identically.

- Digital processing systems do not suffer from the problems above.

- Digital signal processing systems sample the input signal and convert the samples to equivalent digital values. These values are processed and the resulting digital outputs converted back to analogue voltages. This series of discrete voltages is then smoothed to produce the processed analogue output.

- The analogue input signal must be sampled at a frequency which is at least twice as high as its highest frequency component, otherwise 'aliasing' will take place.

1.6 DIGITAL DATA PROCESSING

For the rest of this chapter we will concentrate on the processing of the digital values by the digital data processing unit – this is where the clever bit is done!

So, how does it all work? The digital data processor (Fig. 1.4) is constantly being bombarded with digital values, one following the other at regular intervals. Its job is to output a suitable digital number in response to each digital input. This is something of an achievement as all that the processor has to work with is the current input value and the previous input and output samples. Somehow it has to use these to generate the output value corresponding to the current input value.

The mechanics of what happens is surprisingly simple. First, a number of the previous input and/or output values are stored in special data storage registers, the number stored depending on the nature of the signal processing to be done. Weighted versions of these stored values are then added to (or subtracted from) the current input sample to generate the corresponding output value – the actual algorithm obviously depending on the type of signal processing required. It is this processing algorithm which is at the heart of the whole system – arriving at this can be a *very* complicated business! This is something we will examine in detail in later chapters. Here we will look at some fairly simple examples of processing, just to get a feel for what is involved.

1.7 THE RUNNING AVERAGE FILTER

A good example to start with is the *running (or moving) average filter*. This processing system merely outputs a value which is the average of the current input and a particular number of the *previous* input samples.

As an example, consider a simple running average filter that averages the current input and the *last three* input samples. Let's assume that the sampled input values are as shown in Table 1.1, where T represents the sampling period.

Table 1.1

Time	Input sample
0	2
T	1
2T	4
3T	5
4T	7
5T	10
6T	8
7T	7
8T	4
9T	2

As we need to average the current sample and the previous *three* input samples, the processor will clearly need three registers to store the previous input samples, the contents of these registers being updated every time a new sample is taken. For simplicity, we will assume that these three registers have initially been reset, i.e. they contain the value zero.

The following sequence shows how the first three samples of '2', '1', and '4' are processed:

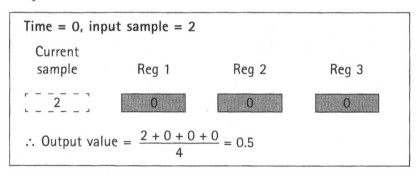

Time = 0, input sample = 2

Current sample — Reg 1 — Reg 2 — Reg 3

2 | 0 | 0 | 0

$$\therefore \text{Output value} = \frac{2 + 0 + 0 + 0}{4} = 0.5$$

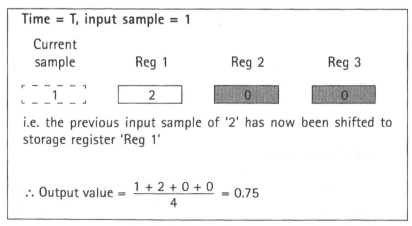

Time = T, input sample = 1

Current sample — Reg 1 — Reg 2 — Reg 3

1 | 2 | 0 | 0

i.e. the previous input sample of '2' has now been shifted to storage register 'Reg 1'

$$\therefore \text{Output value} = \frac{1 + 2 + 0 + 0}{4} = 0.75$$

Time = 2*T*, input sample = 4

Current sample	Reg 1	Reg 2	Reg 3
4	1	2	0

$\therefore$ Output value $= \dfrac{4 + 1 + 2 + 0}{4} = 1.75$

and so on.

Table 1.2 shows all of the output values – check that you agree with them before moving on.

Table 1.2

Time	Input sample	Output sample
0	2	0.5
T	1	0.75
2*T*	4	1.75
3*T*	5	3.00
4*T*	7	4.25
5*T*	10	6.50
6*T*	8	7.50
7*T*	7	8.00
8*T*	4	7.25
9*T*	2	5.25

N.B.1   The first three output values of 0.5, 0.75 and 1.75, represent the initial 'transient', i.e. the part of the output signal where the initial three zeros are being shifted out of the three storage registers. The output values are only valid once these initial zeros have been cleared out of the storage registers.

N.B.2   A running average filter tends to smooth out any rapid changes in a signal and so is a form of lowpass filter.

1.8   REPRESENTATION OF PROCESSING SYSTEMS

The running average filter, just discussed, could be represented by the block diagram shown in Fig. 1.8. Each of the three '*T*' blocks represent a time delay of one sample period, while the 'Σ' box represents the summation of the four values. The '0.25' triangle is an attenuator which ensures that the average of the four values is outputted and not just the sum. So *A* is the current input divided by four, *B* the previous input, again divided by four, *C* the input before that, again divided by four etc. If we catch the system at 6*T* say, then, from Table 1.2, *A* = 8/4, *B* = 10/4, *C* = 7/4 and *D* = 5/4, giving the output of 7.5, i.e. *A* + *B* + *C* + *D*.

N.B.   The division by four could have been done *after* the summation rather than before, and this might seem the obvious thing to do. However, the option used is preferable as it means that, as we are processing smaller numbers, i.e. numbers already divided by four, we can get away with using smaller registers during processing. Here there were only four numbers to be added, but what if

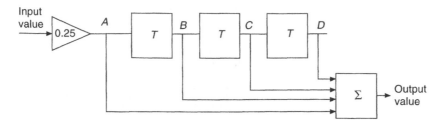

Figure 1.8

there had been a thousand? Dividing *before* addition, rather than after, would clearly makes a huge difference to the size of the registers needed.

1.9 SELF-ASSESSMENT TEST

Calculate the corresponding output samples if the sampled signal, shown in Fig. 1.9, is passed through a running average filter. The filter averages the current input sample and the previous *two* samples. Assume that the two processor storage registers needed are initially reset, i.e. contain the value zero. (The worked solution is given towards the end of the book.)

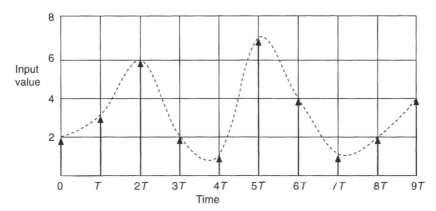

Figure 1.9

1.10 FEEDBACK (OR RECURSIVE) FILTERS

So far we have only met filters which make use of previous *inputs*. There is nothing to stop us from using the previous *outputs* instead – in fact, much more useful filters can be made in this way. A simple example is shown in Fig. 1.10.

Because it is the previous *output* values which are fed back into the system and added to the current input, these filters are called *feedback* or *recursive* filters. Another name very commonly used is *infinite impulse response* filters – the reason for this particular name will become clear later.

As we know, the *T* boxes represent time delays of one sampling period, and so *A* is the previous output and *B* the one before that. It is often useful to think of

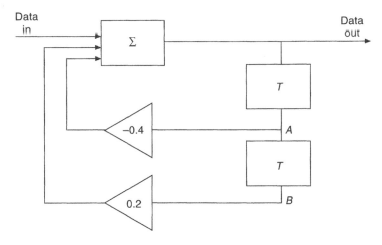

Figure 1.10

these boxes as the storage registers for the previous outputs, with A and B being their contents.

From Fig. 1.10 you should see that:

$$\text{Data out} = \text{Data in} - 0.4A + 0.2B \tag{1.1}$$

This is the simple processing that the digital processor needs to do for every input sample.

Imagine that this particular recursive filter is supplied with the data shown in Table 1.3, and that the two storage registers needed are initially reset.

Table 1.3

Time	Input data	A	B	Output data
0	10	0.0	0.0	10.0
T	15	10.0	0.0	11.0
2T	20	11.0	10.0	17.6
3T	15	17.6	11.0	10.2
4T	8	10.2	17.6	7.5
5T	6	7.5	10.2	5.1
6T	9	5.1	7.5	8.5
7T	0	8.5	5.1	−2.4
8T	0	−2.4	8.5	2.7
9T	0	2.7	−2.4	−1.53

From equation (1.1), as both A and B are initially zero, the first output must be the same as the input value, i.e. '10'.

By the time the second input of 15 is received the previous output of 10 has been shifted into the storage register, appearing as 'A' (Table 1.3). In the meantime, the previous A value (0) has been moved to B, while the previous value of B has been shifted right out of the system and lost, as it is no longer of any use.

So, when time $= T$, we have:

Input data = 15, $A = 10$, $B = 0$

From equation (1.1), the new output value is given by:

$15 - 0.4 \times 10 + 0.2 \times 0 = 11$

In preparation for generating the next output value, the current output of 11 is now shifted to A, the A value of 10 having already been moved to B. The third output value is therefore given by:

$20 - 0.4 \times 11 + 0.2 \times 10 = 17.6$

Before moving on it's best to check through the rest of Table 1.3 for yourself. (Spreadsheets lend themselves well to this application.)

You will notice from Table 1.3 that we are getting outputs *even when the input values are zero* (at times $7T$, $8T$ and $9T$). This makes sense as, at time $7T$, we are pushing the previous output value of 8.5 back through the system to produce the next output of –2.4. This output value is, in turn, fed back into the system, and so on. Theoretically, the output could continue for ever, i.e. even if we put just a single pulse into the system we could get output values, every sampling period, *for an infinite time*. This explains the alternative name of *infinite impulse response* (*IIR*) filter for feedback filters. Note that this persisting output will not happen with processing systems which use only the input samples ('non-recursive') – with these, once the input stops, the output will continue for only a finite time. To be more specific, the output samples will continue for a time of $N \times T$, where N is the number of storage registers. This is why filters which make use of only the *previous* inputs are often called *finite impulse response* (FIR) filters (pronounced 'F-I-R'), for short. A running average filter is therefore an example of an FIR filter. (Yet another name used for this type of filter is the *transversal* filter.)

IIR filters require fewer storage registers than equivalent FIR filters. For example, a particular highpass FIR filter might need 100 registers but an equivalent IIR filter might need as few as three or four. However, I must add a few words of warning here, as there are drawbacks to making use of the previous *outputs*. As with *any* system which uses feedback, we have to be *very* careful during the design as it is possible for the filter to become unstable. In other words, instead of acting as a well-behaved system, processing our signals in the required way, we might find that the output values very rapidly shoot up to the maximum possible and sit there. Another possibility is that the output oscillates between the maximum and minimum values. Not a pretty sight! We will look at this problem in more detail in later chapters.

So far we have looked at systems which make use of either previous inputs or previous outputs only. This restriction is rather artificial as, generally, the most effective DSP systems use both previous inputs *and* previous outputs.

1.11   SELF-ASSESSMENT TEST

As has just been mentioned, DSP systems often make use of previous inputs *and* previous outputs. Figure 1.11 represents such a system.

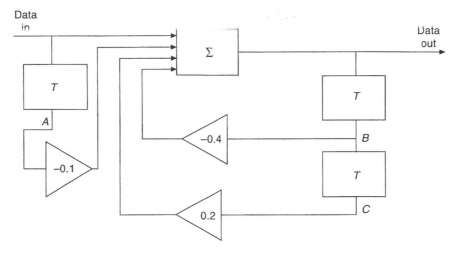

Figure 1.11

(a) Derive the equation which describes the processing performed by the system, i.e. relates the output value to the corresponding input value.

(b) If the input sequence is 3, 5, 6, 6, 8, 10, 0, 0 determine the corresponding output values. Assume that the three storage registers needed (one for the previous input and two for the previous outputs) are initially reset, i.e. contain the value zero.

1.12 CHAPTER SUMMARY

Hopefully, you now have a reasonable understanding of the basics of digital signal processing. You should also realize that this type of signal processing is achieved in a very different way from 'traditional' analogue signal processing. In this chapter we have concentrated on the heart of the DSP system, i.e. the part that processes the digital samples of the original analogue signal. Several processing systems have been analysed. At this stage it will not be clear how these systems are designed to achieve a particular type of signal processing, or even the nature of the signal processing being carried out. This very important aspect will be dealt with in more detail in later chapters. We have met finite impulse response filters (those that make use of previous input samples only, such as running average filters) and also infinite impulse response filters (also called feedback or recursive filters) – these make use of the previous output samples. Although 'IIR' systems generally need fewer storage registers than equivalent 'FIR' systems, IIR systems can be unstable if not designed correctly, while FIR systems will *never* be unstable.

1.13 PROBLEMS

1 Describe, briefly, *four* problems associated with analogue signal processing systems but not with digital signal processing systems.

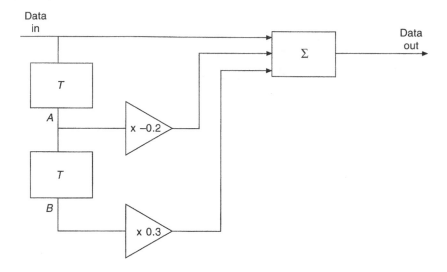

Figure 1.12

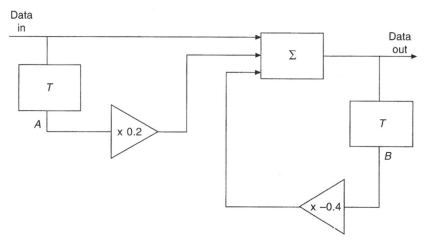

Figure 1.13

2 'An FIR filter generally needs more storage registers than an equivalent IIR filter.' True or false?

3 A running average filter has a frequency response which is similar to that of:

(a) a lowpass filter;

(b) a highpass filter;

(c) a bandpass filter;

(d) a bandstop filter.

4 What are two alternative names for IIR filters?

5 'Once an input has been received, the output from a running average filter can, theoretically, continue for ever.' True or false?

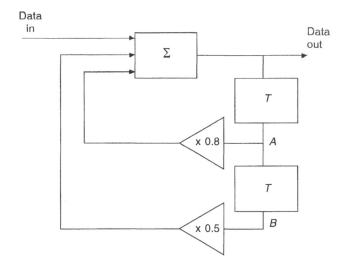

Figure 1.14

6 'An FIR filter can be unstable if not designed correctly.' True or false?

7 A signal is sampled for 4 s by an 8-bit ADC at a sampling frequency of 10 kHz. The samples are stored on a memory chip.

(a) What is the minimum memory required to store all of the samples?

(b) What should be the highest frequency component of the signal?

8 If the sequence 3, 5, 7, 6, 3, 2 enters the processor shown in Fig. 1.12, find the corresponding outputs.

9 The sequence 2, 3, 5, 4 is inputted to the processing system of Fig. 1.13. Find the first *six* outputs.

10 A single input sample of '1', enters the system of Fig. 1.14. Find the first *eight* output values. Is there something slightly worrying about this system?

2 Discrete signals and systems

2.1 CHAPTER PREVIEW

In this chapter we will look more deeply into the nature and properties of discrete signals and systems. We will first look at a way of representing discrete signals in the time domain. You will then be introduced to the *z-transform*, which will allow us to move from the time domain and enter the *z-domain*. This very powerful transform is to discrete signals what the Laplace transform is to continuous signals. It allows us to analyse and design discrete systems much more easily than if we were to remain in the time domain. Recursive and non-recursive filters will be revisited in this new domain. We will also look at how the software package, MATLAB, can be used to help with the analysis of discrete processing systems.

2.2 SIGNAL TYPES

Before we start to look at discrete signals in more detail, it's worth briefly reviewing the different categories of signals.

Continuous-time signals

This is the type of signal produced when we talk or sing, or by a musical instrument. It is 'continuous' in that it is present at all times during the talking or the singing. For example, if a note is played near a microphone then the output voltage from the microphone might vary as shown in Fig. 2.1. Clearly, continuous does *not* mean that it goes on for ever! Other examples would be the output from a signal generator when used to produce sine, square, triangular waves etc. They are strictly called *continuous-time, analogue signals*.

Discrete-time signals

These, on the other hand, are defined at particular or *discrete* instants only – they are usually sampled signals. Some simple examples would be the distance travelled by a car, atmospheric pressure, the temperature of a process, audio signals, *but all recorded at certain times*. The signal is often defined at regular time intervals. For example, atmospheric pressure might be sampled at the same time each day (Fig. 2.2), whereas an audio signal would obviously have to be sampled much

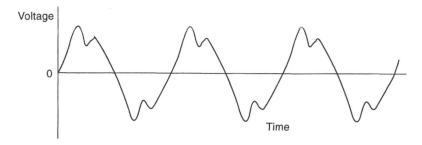

Figure 2.1

more frequently, perhaps every 10 μs, in order to produce a reasonable representation of the signal. (The audio signal could contain frequency components of up to 20 kHz and so, to prevent aliasing, it must be sampled at a minimum of 40 kHz, i.e. at least every 25 μs.)

Digital signals

This term is used to describe a discrete-time signal where the sampled, analogue values have been converted to their equivalent digital values. For example, if the pressure values of Fig. 2.2 are to be automatically processed by a computer then, clearly, they will first need to be changed to equivalent voltage values by means of a suitable transducer. An ADC is then needed to convert the resulting voltages to a series of digital values. It is this series of numbers that constitutes the digital signal.

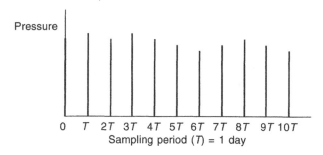

Figure 2.2

2.3 THE REPRESENTATION OF DISCRETE SIGNALS

Before we can start to consider *discrete* signal processing in detail, we need to have some shorthand way of describing our signal. *Continuous* signals can usually be represented by continuous mathematical expressions. For example, if a signal is represented by $y = 3 \sin 4t$, then you will recognize this as representing a sine wave having an angular frequency, ω, of 4 rad/s, and an amplitude of 3. If the signal had been expressed as $y = 3e^{-2t} \sin 4t$, then this is again the same sinusoidal variation but the signal is now decaying exponentially with time. More complicated continuous signals, such as speech, are obviously much more difficult to express mathematically, but you get the point.

Here we are mainly interested in discrete-time signals, and these are very different, in that *they are only defined at particular times*. Clearly, some other way of representing them is needed. If we had to describe the regularly sampled signal shown in Fig. 2.3, for example, then a sensible way would be as '$x[n] = 1$, 3, 2, 0, 2, 1'. This method is fine and is commonly used. $x[n]$ indicates a sequence of values where n refers to the nth position in the sequence. So, for this particular sequence, $x[0] = 1$, $x[1] = 3$, $x[2] = 2$, and so on.

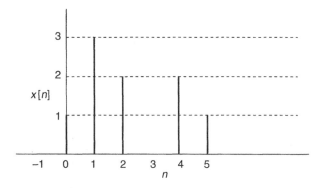

Figure 2.3

Some simpler signals can be expressed more succinctly. For example, how about the *sampled* or *discrete unit step* shown in Fig. 2.4? Using the normal

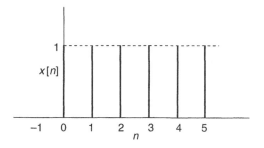

Figure 2.4

convention, we *could* represent this as $x[n] = 1, 1, 1, 1, 1, 1, \ldots$. However, as the '1' continues indefinitely, a neater way would be as:

'$x[n] = 1$ for $n \geq 0$, else $x[n] = 0$'

In other words, $x[n] = 1$ for n is zero or positive and is zero for all negative n values.

Because of its obvious similarity to the *continuous* unit step, which is usually indicated by $u(t)$, the *discrete* unit step is often represented by the special symbol $u[n]$:

$u[n] = 1, 1, 1, 1, 1, \ldots$

Another simple *but extremely important* discrete signal is shown in Fig. 2.5. This is the *unit sample function* or *the unit sample sequence*, i.e. a single unit pulse at $t = 0$. (The importance of this signal will be explained in a later chapter.)

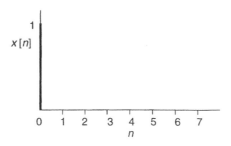

Figure 2.5

This sequence could be represented by $x[n] = 1, 0, 0, 0, 0, \ldots$ or, more simply, by:

'$x[n] = 1$ for $n = 0$, else $x[n] = 0$'

This particular function is equivalent to the *unit impulse* or the *Dirac delta function*, $\delta(t)$, from the 'continuous' world and is often called the *Kronecker delta function* or the *delta sequence*. Because of this similarity it is very sensibly represented by the symbol $\delta[n]$:

$\delta[n] = 1, 0, 0, 0, \ldots$

A *delayed* unit sample/delta sequence, for example, one delayed by two sample periods (Fig. 2.6) is represented by $\delta[n - 2]$ where the '–2' indicates a delay of two sample periods.

$\delta[n - 2] = 0, 0, 1, 0, 0, \ldots$

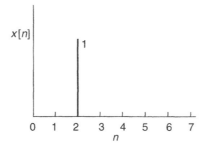

Figure 2.6

Beware! It might be tempting to use $\delta[n + 2]$ instead of $\delta[n - 2]$, and this mistake is sometimes made. To convince yourself that this is *not* correct, you need to start by looking at the original, undelayed sequence. Here it is the zeroth term ($n = 0$) which is 1, i.e. $\delta[0] = 1$.

Now think about the delayed sequence and look where the '1' is. Clearly, this is defined by $n = 2$. So, if we now substitute $n = 2$ into $\delta[n - 2]$, then we get $\delta[2 - 2]$, i.e. $\delta[0]$. So, once again, $\delta[0] = 1$. On the other hand if, in error, we had used '$\delta[n + 2]$' to describe the delayed delta sequence, then this is really stating that $\delta[4] = 1$, i.e. that it is the fourth term in the original, undelayed delta sequence which is 1. This is clearly nonsense. Convinced?

Question

What shorthand symbol could you use to represent the discrete signal shown in Fig. 2.7?

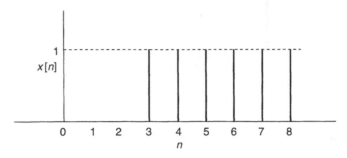

Figure 2.7

Answer

Hopefully your answer is '$u[n - 3]$', i.e. the signal is a discrete unit step, $u[n]$, but delayed by three sampling periods.

Example 2.1

An exponential signal, $x(t) = 4e^{-2t}$, is sampled at a frequency of 10 Hz, beginning at time $t = 0$.

(a)   Express the sampled sequence, $x[n]$, up to the fifth term.

(b)   If the signal is delayed by 0.2 s, what would now be the first five terms?

(c)   What would be the resulting sequence if $x[n]$ and the delayed sequence of (b) above are added together?

Solution

(a)   As the sampling frequency is 10 Hz, then the sampling period is 0.1 s. It follows that the value of the signal needs to be calculated every 0.1 s, i.e.:

Time (s)	0	0.1	0.2	0.3	0.4	0.5
$x(t) = 4e^{-2t}$	4.00	3.27	2.68	2.20	1.80	1.47

   $\therefore x[n] = 4.00, 3.27, 2.68, 2.20, 1.80, 1.47, \ldots$

(b)   If the signal is delayed by 0.2 s, i.e. two samples, then the first two samples will be 0,

   $\therefore x[n - 2] = 0, 0, 4.00, 3.27, 2.68 \ldots$

(c)   As $x[n] = 4.00, 3.27, 2.68, 2.20, 1.80, 1.47, \ldots$ and $x[n - 2] = 0, 0, 4.00, 3.27, 2.68$

then

$x[n] + x[n - 2] = 4.00, 3.27, 6.68, 5.47, 4.48, \ldots$ (first five terms only)

2.4   SELF-ASSESSMENT TEST

1 A signal, given by $x(t) = 2 \cos(3\pi t)$, is sampled at a frequency of 20 Hz, starting at time $t = 0$.

   (a) Is the signal sampled frequently enough? (Hint: 'aliasing')

   (b) Find the first six samples of the sequence.

   (c) Given that the sampled signal is represented by $x[n]$, how could the above signal, *delayed by four sample periods,* be represented?

2 Two sequences, $x[n]$ and $w[n]$, are given by 2, 2, 3, 1 and −1, −2, 1, −4 respectively. Find the sequences corresponding to:

   (a) $x[n] + w[n]$;

   (b) $x[n] + 3(w[n - 2])$;

   (c) $x[n] - 0.5(x[n - 1])$.

3 Figure 2.8 shows a simple digital signal processor, the T block representing a signal delay of one sampling period. The input sequence and the weighted, delayed sequence at A are added to produce the output sequence $y[n]$.

   If the input sequence is $x[n] = 2, 1, 3, -1$, find:

   (a) the sequence at A;

   (b) the output sequence, $y[n]$.

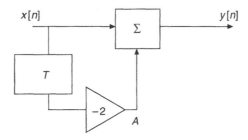

Figure 2.8

2.5   RECAP

- A sampled signal can be represented by $x[n] = 3, 5, 7, 2, 1, \ldots$ for example, where 3 is the sample value at time $= 0$, 5 at time $= T$ (one sampling period), etc. If this signal is delayed by, say, two sampling periods, then this would be shown as $x[n - 2] = 0, 0, 3, 5, 7, 2, 1, \ldots$

- Important discrete signals are the discrete unit step $u[n]$ (1, 1, 1, 1, . . .) and the unit sample sequence, or delta sequence, $\delta[n]$ (1, 0, 0, 0, . . .).

- Sequences can be delayed, scaled, added together, subtracted from each other, etc.

2.6 THE *z*-TRANSFORM

So far we have looked at signals and systems in the time domain. Analysing discrete processing systems in the time domain can be *extremely* difficult. You are probably very familiar with the Laplace transform and the concepts of the complex frequency *s* and the '*s*-domain'. If so you will appreciate that using the Laplace transform, to work in the *s*-domain, makes the analysis and design of *continuous* systems *much* easier than trying to do this in the time domain. Working in the time domain can involve much unpleasantness such as having to solve complicated differential equations and also dealing with the process of convolution. These are activities to be avoided if at all possible!

In a very similar way the *z-transform* will make our task much easier when we have to analyse and design *discrete* processing systems. It takes a process, which can be extremely complicated, and converts it to one that is surprisingly simple – the beauty of the simplification is always impressive to me!

Consider a general sequence

$x[n] = x[0], x[1], x[2], x[3], . . .$

The *z*-transform, $X(z)$, of $x[n]$ is defined as:

$$X(z) = \sum_{n=0}^{\infty} x[n]z^{-n}$$

In other words, we sum the expression, $x[n]z^{-n}$ for *n* values between $n = 0$ and ∞. (The origin of this transform will be explained in the next chapter.)

Expanding the summation makes the definition easier to understand, i.e.:

$$X(z) = x[0]z^0 + x[1]z^{-1} + x[2]z^{-2} + x[3]z^{-3} + . . .$$

As an example of the transformation, let's assume that we have a discrete signal, $x[n]$, given by $x[n] = 3, 2, 5, 8, 4$. The *z*-transform of this finite sequence can then be expressed as:

$X(z) = 3 + 2z^{-1} + 5z^{-2} + 8z^{-3} + 4z^{-4}$

Very simply, we can think of the 'z^{-n}' term as indicating a time delay of *n* sampling periods. For example, the sample of value '3' arrives first at $t = 0$, i.e. with no delay ($n = 0$ and $z^0 = 1$). This is followed, one sampling period later, by the '2', and so this is tagged with z^{-1} to indicate its position in the sequence. Similarly the value 5 is tagged with z^{-2} to indicate the delay of two sampling periods. (There is more to the z operator than this, but this is enough to know for now.)

Example 2.2_____

Transform the finite sequence $x[n] = 2, 4, 3, 0, 6, 0, 1, 2$ from the time domain into the z-domain, i.e. find its z-transform.

Solution

From the definition of the z-transform:

$$X(z) = 2 + 4z^{-1} + 3z^{-2} + 0z^{-3} + 6z^{-4} + 0z^{-5} + z^{-6} + 2z^{-7}$$

or

$$X(z) = 2 + 4z^{-1} + 3z^{-2} + 6z^{-4} + z^{-6} + 2z^{-7}$$

Example 2.3_____

Find the z-transform for the unit sample function, i.e. $\delta[n] = 1, 0, 0, 0, \ldots$

Solution

This one is particularly simple – it's just $\Delta(z) = 1$:

$$1z^0 + 0z^{-1} + 0z^{-2} + \ldots$$

If the sequence had been a unit sample function delayed by three sample periods, for example, i.e. $x[n] = 0, 0, 0, 1, 0, \ldots$ then $X(z) = z^{-3}$. Alternatively, this delayed, unit sample function could be represented very neatly by $z^{-3}\Delta(z)$.

Example 2.4_____

Find the z-transform for the discrete unit step, i.e. $u[n] = 1, 1, 1, 1, 1, 1, \ldots$

Solution

$$U(z) = 1 + z^{-1} + z^{-2} + z^{-3} + z^{-4} + z^{-5}\ldots$$

N.B.   The z-transform of the unit step can be expressed in the 'open form' shown above, and also in an alternative 'closed form'.

The equivalent closed form is

$$U(z) = \frac{z}{z-1}$$

It is not obvious that the open and closed forms are equivalent, but it's fairly easy to show that they are.

First divide the top and bottom by z, to give

$$U(z) = \frac{1}{1 - z^{-1}}$$

The next stage is to carry out a polynomial 'long division':

$$
\begin{array}{r}
1 + z^{-1} + z^{-2} + \dots \\
1 - z^{-1} \overline{)1 + 0 \quad + 0 \quad + 0} \\
\underline{1 - z^{-1}} \\
z^{-1} + 0 \\
\underline{z^{-1} - z^{-2}} \\
z^{-2} + 0 \\
\underline{z^{-2} - z^{-3}} \\
z^{-3}
\end{array}
$$

This gives, $1 + z^{-1} + z^{-2} + \dots$, showing that the open and closed two forms *are* equivalent.

2.7   *z*-TRANSFORM TABLES

We have obtained the z-transforms for two important functions, the unit sample function and the unit step – others are rather more difficult to derive from first principles. However, as with Laplace transforms, tables of z-transforms are available. A table showing both the Laplace and z-transforms for several functions is given in Appendix A. Note that the z-transforms for the unit sample function and the discrete unit step agree with our expressions obtained earlier – which is encouraging!

2.8   SELF-ASSESSMENT TEST

1   By applying the definition of the z-transform, find the z-transforms for the following sequences:

(a) 1, 2, 4, 0, 6;

(b) $x[n] = n/(n + 1)^2$ for $0 \le n \le 3$;

(c) the sequence given in (a) above but delayed by two sample periods;

(d) $x[n] = e^{-0.2n}$ for $0 \le n \le 3$.

2   Given that the sampling frequency is 10 Hz, use the z-transform table of Appendix A to find the z-transforms for the following functions of time, t:

(a) $3e^{-2t}$;   (b) $2 \sin 3t$;   (c) te^{-2t};   (d) t (unit ramp).

2.9   THE TRANSFER FUNCTION FOR A DISCRETE SYSTEM

As you are probably aware, if $X(s)$ is the Laplace transform of the input to a continuous system and $Y(s)$ the Laplace transform of the corresponding output (Fig. 2.9), then the system transfer function, $T(s)$, is defined as:

$$T(s) = \frac{Y(s)}{X(s)}$$

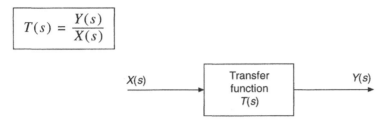

$X(s) \longrightarrow$ | Transfer function $T(s)$ | $\longrightarrow Y(s)$

Figure 2.9

In a similar way, if the z-transform of the input sequence to a discrete system is $X(z)$ and the corresponding output sequence is $Y(z)$ (Fig. 2.10), then the transfer function, $T(z)$, of the system is defined as:

$$T(z) = \frac{Y(z)}{X(z)}$$

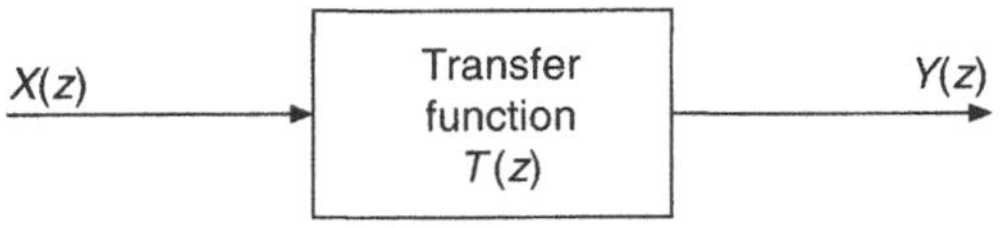

$X(z) \longrightarrow$ | Transfer function $T(z)$ | $\longrightarrow Y(z)$

Figure 2.10

The transfer function for a system, continuous or discrete, is an *extremely* important expression as it allows the response to any input sequence to be derived.

Example 2.5

If the input to a system, having a transfer function of $(1 + 2z^{-1} - z^{-2})$, is a discrete unit step, find the first four terms of the output sequence.

Solution

$$T(z) = \frac{Y(z)}{X(z)}$$

$\therefore\ Y(z) = T(z)\,X(z)$

$\therefore\ Y(z) = (1 + 2z^{-1} - z^{-2})(1 + z^{-1} + z^{-2} + z^{-3})$

(As only the first four terms of the output sequence are required then we need only show the first four of the input sequence.)

$\therefore\ Y(z) = 1 + z^{-1} + z^{-2} + z^{-3} + 2z^{-1} + 2z^{-2} + 2z^{-3} - z^{-2} - z^{-3} \ldots$

$\therefore\ Y(z) = 1 + 3z^{-1} + 2z^{-2} + 2z^{-3} + \ldots$

i.e. the first four terms of the output sequence are 1, 3, 2, 2.

N.B. By working in the z-domain, finding the system response was fairly simple – it is reduced to the multiplication of two polynomial functions. To have found the response in the time domain would have been *much* more difficult.

Example 2.6

When a discrete signal $1, -2$ is inputted to a processing system the output sequence is $1, -5, 8, -4$.

(a)   Derive the transfer function for the system.

(b)   Find the first three terms of the output sequence in response to the finite input sequence of $2, 2, 1$.

Solution

(a) $$T(z) = \frac{Y(z)}{X(z)}$$

$$\therefore T(z) = \frac{1 - 5z^{-1} + 8z^{-2} - 4z^{-3}}{1 - 2z^{-1}}$$

Carrying out a 'long division':

$$
\begin{array}{r}
1 - 3z^{-1} + 2z^{-2} \\
1 - 2z^{-1} \overline{)\,1 - 5z^{-1} + 8z^{-2} - 4z^{-3}} \\
\underline{1 - 2z^{-1}} \\
-3z^{-1} + 8z^{-2} \\
\underline{-3z^{-1} + 6z^{-2}} \\
2z^{-2} - 4z^{-3} \\
\underline{2z^{2} - 4z^{-3}}
\end{array}
$$

i.e. $T(z) = 1 - 3z^{-1} + 2z^{-2}$

(b) $$T(z) = \frac{Y(z)}{X(z)}$$

$\therefore Y(z) = T(z)X(z)$

$\therefore Y(z) = (1 - 3z^{-1} + 2z^{-2})(2 + 2z^{-1} + z^{-2})$

$\therefore Y(z) = 2 + 2z^{-1} + z^{-2} - 6z^{-1} - 6z^{-2} + 4z^{-2}$

(All terms above z^{-2} are ignored as only the first three terms of the output sequence are required.)

$\therefore Y(z) = 2 - 4z^{-1} - z^{-2}$

i.e. the first three terms of the output sequence are $2, -4, -1$.

Example 2.7_____

A processor has a transfer function, $T(z)$, given by $T(z) = 0.4/(z - 0.5)$. Find the response to a discrete unit step (first four terms only).

Solution

We could solve this one in a similar way to the previous example, i.e. by first dividing 0.4 by $(z - 0.5)$ to get the first four terms of the open form of the transfer function, and then multiplying by $1 + z^{-1} + z^{-2} + z^{-3}$. If you do this you should find that the output sequence is 0, 0.4, 0.6, 0.7, to four terms.

However, there is an alternative approach, which avoids the need to do the division. This method makes use of the z-transform tables to find the inverse z-transform of $Y(z)$.

Using the closed form, $z/(z - 1)$, for $u(z)$, we get:

$$Y(z) = \left(\frac{z}{z - 1} \right) \left(\frac{0.4}{z - 0.5} \right)$$

We now need to find the inverse z-transform. The first thing we must do is use the method of *partial fractions* to change the format of $Y(z)$ into functions which appear in the transform tables. (You might have used this technique before to find inverse Laplace transforms.) For reasons which will be explained later, it's best to find the partial fractions for $Y(z)/z$ rather than $Y(z)$. This is *not* essential but it does make the process much easier. Once we have found the partial fractions for $Y(z)/z$ all we then need to do is multiply them by z, to get $Y(z)$:

$$\frac{Y(z)}{z} = \frac{0.4}{(z - 1)(z - 0.5)}$$

$$\therefore \frac{Y(z)}{z} = \frac{0.4}{(z - 1)(z - 0.5)} = \frac{A}{(z - 1)} + \frac{B}{(z - 0.5)}$$

Using the 'cover-up rule', or any other method, you should find that $A = 0.8$ and $B = -0.8$.

$$\therefore \frac{Y(z)}{z} = \frac{0.8}{(z - 1)} - \frac{0.8}{(z - 0.5)}$$

$$\text{or } Y(z) = \frac{0.8z}{(z - 1)} - \frac{0.8z}{(z - 0.5)}$$

Now $z/(z - 1)$ is the z-transform of the unit step, and so $0.8z/(z - 1)$ must have the inverse z-transform of 0.8, 0.8, 0.8, 0.8, ...

The inverse transform of $0.8z/(z - 0.5)$ is slightly less obvious. Looking at the tables, we find that the one we want is $z/(z - e^{-aT})$.

Comparing with our expression, we see that $0.5 \equiv e^{-aT}$. The function $z/(z - e^{-aT})$ inverse transforms to e^{-at} or, *as a sampled signal*, to e^{-anT}. As $0.5 \equiv e^{-aT}$, then our function must inverse transform to $0.8(0.5^n)$, i.e. to $0.8(0.5^0, 0.5^1, 0.5^2, 0.5^3, \ldots)$ or 0.8, 0.4, 0.2, 0.1, to four terms.

Subtracting this sequence from the '0.8' step, i.e. 0.8, 0.8, 0.8, . . . , gives the sequence of 0, 0.4, 0.6, 0.7, . . . to four terms.

N.B.1   If we had found the partial fractions for $Y(z)$, rather than $Y(z)/z$, then these would have had the form of $b/(z - 1)$, where b is a number. However, this expression does not appear in the z-transform tables. This isn't an insurmountable problem but it is easier to find the partial fractions for $Y(z)/z$ and then multiply the resulting partial fractions by z to give a recognizable form.

N.B.2   It would not have been worth solving Example 2.5 in this way, as it would have involved much more work than the method used. However, it is worth ploughing through it, just to demonstrate the technique once again.

The transfer function here is $1 + 2z^{-1} - z^{-2}$ and the input a unit step:

$$\therefore Y(z) = \frac{z}{z-1} (1 + 2z^{-1} - z^{-2})$$

or

$$Y(z) = \frac{z}{z-1} \frac{z^2 + 2z - 1}{z^2} = \frac{z^2 + 2z - 1}{z(z-1)} \text{ (using positive indices)}$$

i.e.

$$\frac{Y(z)}{z} = \frac{z^2 + 2z - 1}{z^2(z-1)} = \frac{A}{z-1} + \frac{B}{z} + \frac{C}{z^2}$$

You should find that this gives $A = 2$, $B = -1$ and $C = 1$.

$$\therefore \frac{Y(z)}{z} = \frac{2}{z-1} - \frac{1}{z} + \frac{1}{z^2}$$

or

$$Y(z) = \frac{2z}{z-1} - 1 + \frac{1}{z}$$

We could refer to the z-transform tables at this point but, by now, you should recognize $2z/(z-1)$ as the z-transform for a discrete 'double' step (2, 2, 2, 2, 2, . . .), '−1' for the unit sample sequence, but inverted, i.e. −1, 0, 0, . . . and $1/z$ or z^{-1}, for the unit sample sequence, but delayed by one sampling period, i.e. 0, 1, 0, 0, ...

Adding these three sequences together gives us 1, 3, 2, 2, to four terms, which is what we found using the alternative 'long division' method.

2.10   SELF-ASSESSMENT TEST

1   For a particular digital processing system, inputting the finite sequence $x[n]$ = (1, −1) results in the finite output sequence of (1, 2, −1, −2). Determine the z-transforms for the two signals and show that the transfer function for the processing system is given by $T(z) = 1 + 3z^{-1} + 2z^{-2}$.

2   A digital processing system has the transfer function of $(1 + 3z^{-1})$. Determine the output sequences given that the finite sequences

(a) 1, 0, 2 and (b) 2, 3, −1 are inputted to the system.

3   A digital filter has a transfer function of $z/(z - 0.6)$. Use z-transform tables to find its response to (a) a unit sample function and (b) a discrete unit step (first three terms only).

2.11   MATLAB AND SIGNALS AND SYSTEMS

In the previous section we looked at manual methods for finding the response of various processing systems to different discrete inputs. Although it's important to understand the process, *in practice* it would be unusual to do this manually. This is because there are plenty of software packages available that will do the work for us. One such package is MATLAB. This is an excellent CAD tool which has many applications ranging from the solution of equations and the plotting of graphs to the complete design of a digital filter. The specialist MATLAB 'toolbox', 'Signals and Systems', will be used throughout the remainder of this book for the analysis and design of DSP systems. Listings of some of the MATLAB programs used are included. Clearly, it will be extremely useful if you have access to this package.

Figure 2.11 shows the MATLAB display corresponding to Example 2.5, i.e. for $T(z) = 1 + 2z^{-1} - z^{-2}$ and an input of a discrete unit step.

N.B.   The output is displayed as a series of steps rather than the correct output of a sequence of pulses. However, in spite of this, it can be seen from the display that the output sequence is 1, 3, 2, 2, ..., which agrees with our values obtained earlier. (It *is* possible to display proper pulses – we will look at this improvement later.)

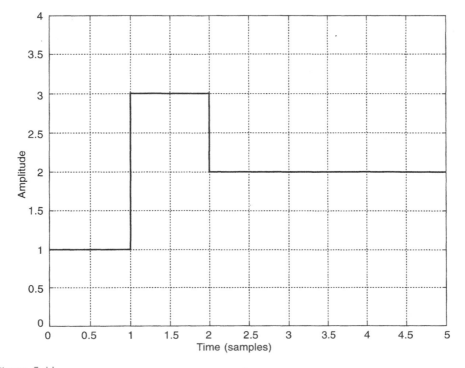

Figure 2.11

The program used to produce the display of Fig. 2.11 is as follows:

```
numz=[1 2 -1];      % coefficients of numerator, z² + 2z
                    % - 1 of transfer function
                    % (N.B. +ve indices must be used, and
                    % so numerator and
                    % denominator must be multiplied by z²)
denz=[1 0 0];       % coefficients of denominator (z² +
                    % 0z + 0)
dstep(numz,denz)    % displays step response      .
axis([0 5 0 4])     % overrides default axis with 'x'
                    % axis 0 to 5 and 'y' axis 0 to 4
                    % (N.B. This is not always necessary)
grid                % adds grid
```

When the inputs are discrete steps or unit sample functions it is quite easy to display the outputs, as 'dstep' and 'dimpulse' (or 'impz' in some versions) are standard MATLAB functions ('dimpulse (numz,denz)' would have generated the response to a unit sample function). However, if some general input sequence is used, such as the one in Example 2.6, then it is slightly more complicated. The program needed for this particular example, i.e. when $T(z) = 1 - 3z^{-1} + 2z^{-2}$ and $x[n] = (2, 2, 1)$ is:

```
numz=[1 -3 2];       % Transfer function numerator
                     % coefficients (z² - 3z + 2)
denz=[1 0 0];        % denominator coefficients (z² + 0z + 0)
x=[2 2 1];           % input sequence
dlsim(numz,denz,x);  % displays output (See MATLAB manual for
                     % further details of this function)
grid                 % adds grid
```

The MATLAB display is shown in Fig. 2.12.

N.B.1   If you have access to MATLAB then it would be a good idea to use it to check the output sequences obtained in the other worked examples and also the self-assessment exercises.

N.B.2   MATLAB displays can be improved by adding labels to the axes, titles etc. You will need to read the MATLAB user manual to find out about these and other refinements.

2.12 RECAP

• The z-transform can greatly simplify the design and analysis of discrete processing systems.

• The z-transform, $X(z)$, for a sequence, $x[n]$, is given by:

$$X(z) = \sum_{n=0}^{\infty} x[n]z^{-n}$$

or

$$X(z) = x[0]z^0 + x[1]z^{-1} + x[2]z^{-2} + x[3]z^{-3} + \ldots.$$

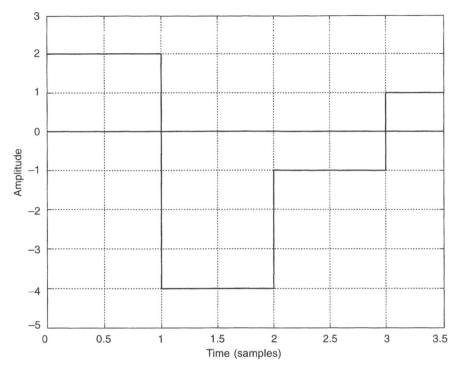

Figure 2.12

- z^{-n} can be thought of as a 'tag' to a sample, the '$-n$' indicating a delay of n sampling periods.

- Important z-transforms are for the discrete unit step, i.e. $U(z) = z/(z - 1)$ and the unit sample function, $\Delta(z) = 1$.

- The transfer function for a discrete system is given by $T(z) = Y(z)/X(z)$, where $Y(z)$ and $X(z)$ are the z-transforms of the output and input sequences respectively.

- MATLAB is a powerful software tool which can be used, amongst many other things, to analyse and design DSP systems.

2.13 DIGITAL SIGNAL PROCESSORS AND THE z-DOMAIN

In Chapter 1 you were introduced to different types of digital signal processors. You will remember that digital filters can be broadly classified as *recursive* or *non-recursive*, i.e. filters that feed back previous output samples to produce the current output sample (recursive), and those that don't. An alternative name for the recursive filter is the infinite impulse response filter (IIR). This is because the filter response to a unit sample sequence is *usually* an infinite sequence – normally decaying with time. On the other hand, the response of a non-recursive filter to a unit sample sequence is always a finite sequence – hence the alternative name of finite impulse response filter (FIR).

Also in Chapter 1, as an introduction to filter action, the response of filters to various input sequences was found by using tables to laboriously add up sample values at various points in the system. There we were working in the time domain. A more realistic approach, still in the time domain, would have been to use the process of 'convolution' – a technique you will probably have met when dealing with continuous signals. If so, you will appreciate that this technique has its own complications. However, we have seen that, by moving from the time domain into the z-domain, the system response can be found much more simply.

As an example, consider the simple discrete processing system shown in Fig. 2.13. As only the previous input samples are used, along with the current input, you will recognize this as an FIR or non-recursive filter. As the input and output sequences are represented by $x[n]$ and $y[n]$ respectively, and the time delay of one sample period is depicted by a 'T' box (where T is the sampling period), then this is obviously the *time domain* model. The processing being carried out is fairly simple, in that the input sequence, delayed by one sampling period and multiplied by a factor of –0.4, is added to the undelayed input sequence to generate the output sequence, $y[n]$. This can be expressed mathematically as:

$$y[n] = x[n] - 0.4x[n - 1] \tag{2.1}$$

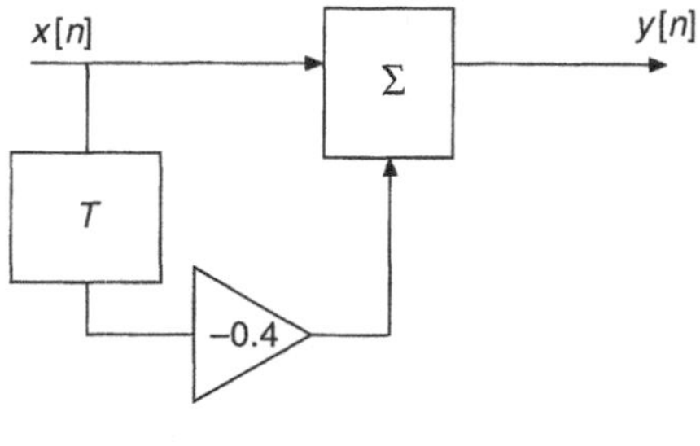

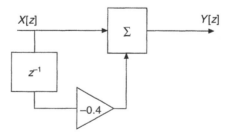

Figure 2.13

However, we have chosen to abandon the time domain and work in the z-domain and so the first thing we will need to do is convert our processing system of Fig. 2.13 to its z-domain version. This is shown in Fig. 2.14.

Figure 2.14

This is a very simple conversion to carry out. The input and output sequences, $x[n]$ and $y[n]$, of Fig. 2.13, have just been replaced by their z-transforms, and the time delay of T is now represented by z^{-1}, the usual way of indicating a time delay of one sampling period in the z-domain.

The mathematical representation of the 'z-domain' processor now becomes:

$$Y(z) = X(z) - 0.4X(z)z^{-1}$$

or

$$Y(z) = X(z)(1 - 0.4z^{-1})$$

$$\therefore \frac{Y(z)}{X(z)} = T(z) = (1 - 0.4z^{-1})$$

i.e. we can find the transfer function of the processor quite easily. Armed with this we should be able to predict the response of the system to any input, either manually, or with the help of MATLAB.

N.B.   The z-domain transfer function could also have been found by just converting equation (2.1) to its z-domain equivalent, i.e. $Y(z) = X(z) - 0.4X(z)z^{-1}$, and so on ($X(z)z^{-1}$ represents the input sequence, delayed by one sampling period).

2.14   FIR FILTERS AND THE z-DOMAIN

The processor we have just looked at is a simple form of a finite impulse response, or non-recursive filter. The z-domain model of a general FIR filter is shown in Fig. 2.15.

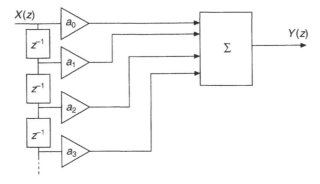

Figure 2.15

Each z^{-1} box indicates a further delay of one sampling period. For example, the input to the amplifier a_3 has been delayed by three sample periods and so the signal leaving the amplifier can be represented by $a_3X(z)z^{-3}$.

The processor output is the sum of the input sequence and the various delayed sequences and is given by:

$$Y(z) = a_0X(z) + a_1X(z)z^{-1} + a_2X(z)z^{-2} + a_3X(z)z^{-3} + \ldots$$

and so:

$$\boxed{T(z) = a_0 + a_1z^{-1} + a_2z^{-2} + a_3z^{-3} \ldots}$$

(2.2)

2.15   IIR FILTERS AND THE *z*-DOMAIN

These, of course, are characterized by a feedback loop. Weighted versions of delayed inputs *and outputs* are added to the input sample to generate the current output sample. Figure 2.16 shows the *z*-domain model of a simple example of this type of filter which adds only previous outputs to the current input.

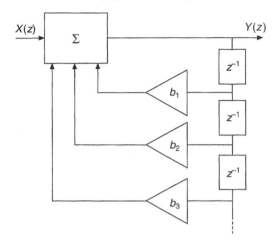

Figure 2.16

Example 2.8_____

What is the transfer function of the IIR filter shown in Fig. 2.16?

Solution

This one is slightly more difficult to derive than that for the FIR filter. However, as before, we first need to find $Y(z)$.

From Fig. 2.16:

$$Y(z) = X(z) + b_1 Y(z)z^{-1} + b_2 Y(z)z^{-2} + b_3 Y(z)z^{-3}$$

$$\therefore Y(z) = X(z) + Y(z)(b_1 z^{-1} + b_2 z^{-2} + b_3 z^{-3})$$

$$\therefore Y(z)(1 - b_1 z^{-1} - b_2 z^{-2} - b_3 z^{-3}) = X(z)$$

$$\boxed{\therefore T(z) = \frac{1}{1 - b_1 z^{-1} - b_2 z^{-2} - b_3 z^{-3}}}$$

Figure 2.17 shows a more general IIR filter, making use of both the previous outputs *and* inputs. This arrangement provides greater versatility.

It shouldn't be too difficult to show that the transfer function for this general recursive filter is given by:

$$\boxed{T(z) = \frac{a_0 + a_1 z^{-1} + a_2 z^{-2} + a_3 z^{-3} + \dots}{1 - b_1 z^{-1} - b_2 z^{-2} - b_3 z^{-3}}} \tag{2.3}$$

Try it!

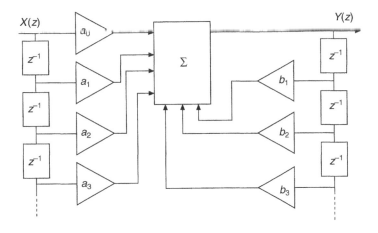

Figure 2.17

Reminder

- Recursive (IIR) filters can be used to achieve the same processing as almost all non-recursive (FIR) filters but have the advantage of needing fewer multipliers/coefficients and so are simpler in terms of hardware.

- As with any system that uses feedback, a recursive filter can be unstable – this is not the case with FIR filters – these are always stable. (We will find later that IIR filters also have a 'phase problem' compared to FIR filters.)

Example 2.9_____

An FIR filter has three coefficients, $a_0 = 0.5$, $a_1 = -0.8$ and $a_2 = 0.4$. Derive the output sequence corresponding to the finite input sequence of 1, 5.

Solution

We could derive this from first principles, by first drawing the z-domain diagram. However, by inspection of the transfer function for an FIR filter given earlier (equation (2.2)):

$$T(z) = 0.5 - 0.8z^{-1} + 0.4z^{-2}$$

Also, as $x[n] = 1, 5$:

$$X(z) = 1 + 5z^{-1}$$

$$\therefore \quad Y(z) = (1 + 5z^{-1})(0.5 - 0.8z^{-1} + 0.4z^{-2})$$

which gives:

$$Y(z) = 0.5 + 1.7z^{-1} - 3.6z^{-2} + 2z^{-3}$$

So the output sequence is 0.5, 1.7, –3.6, 2.
The MATLAB output display is shown in Fig. 2.18.

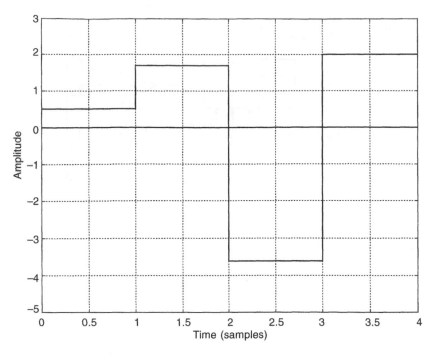

Figure 2.18

Example 2.10

A filter has a transfer function given by

$$T(z) = \frac{1 - 0.4z^{-1}}{1 + 0.2z^{-2}}$$

(a)   Draw the z-domain version of the block diagram for the filter.

(b)   Derive an expression for the output sequence $y[n]$, in terms of the input sequence, $x[n]$, and delayed input and output sequences.

(c)   Find the unit sample response for the filter (first four terms only).

Solution

(a)   This is clearly a recursive filter. By comparison with the general expression given earlier (equation (2.3)), this filter has coefficients of $a_0 = 1$, $a_1 = -0.4$, $b_1 = 0$, $b_2 = -0.2$ and so the z-domain block diagram is as shown in Fig. 2.19.

N.B.   The two 'z^{-1}' boxes, used to feed back the output sequence, could be replaced by a single 'z^{-2}' box.

(b) $$\frac{Y(z)}{X(z)} = \frac{1 - 0.4z^{-1}}{1 + 0.2z^{-2}}$$

$$\therefore Y(z)(1 + 0.2z^{-2}) = X(z)(1 - 0.4z^{-1})$$

$$\therefore Y(z) = X(z) - 0.4X(z)z^{-1} - 0.2Y(z)z^{-2}$$

Changing to the time domain:

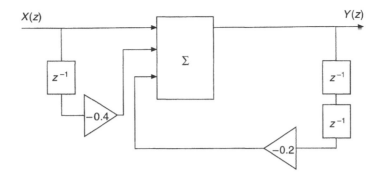

Figure 2.19

$$y[n] = x[n] - 0.4x[n{-}1] - 0.2y[n{-}2]$$

(c) $$T(z) = \dfrac{1 - 0.4z^{-1}}{1 + 0.2z^{-2}}$$

Carrying out the polynomial division gives $T(z) = 1 - 0.4z^{-1} - 0.2z^{-2}$, to three terms. Check this for yourself – note that you will need to divide $1 - 0.4z^{-1}$ by $1 + 0z^{-1} + 0.2z^{-2}$.

As the z-transform of the input is just '1', the z-transform of the output sequence will be exactly the same as the transfer function, i.e. $Y(z) = 1 - 0.4z^{-1} - 0.2z^{-2}$ and so the first three terms of the output sequence will be 1, –0.4, –0.2.

The MATLAB output display is shown in Fig. 2.20.

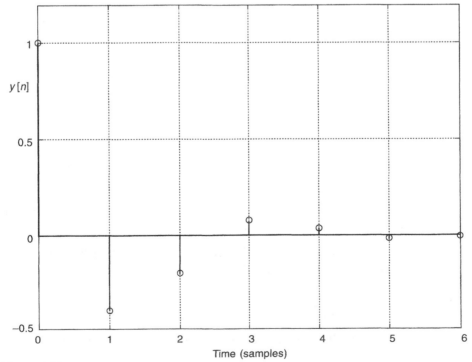

Figure 2.20

N.B.   The sampled output signal is shown in Fig. 2.20 as pulses, rather than 'steps'. This improved
display is achieved by using the MATLAB 'stem' function. The program is as follows:

```
numz=[1 -0.4 0];            % numerator coefficients after
                            % multiplying by z²
denz=[1 0 0.2];             % denominator coefficients after
                            % multiplying by z²
n=(0:6);                    % 6 samples to be displayed
[y x]= dimpulse(numz,denz,n);   % evaluate output values (y)
stem(n,y)                   % display output sequence as 'stems'
grid                        % add grid to display
```

2.16   SELF-ASSESSMENT TEST

1   Derive the transfer function for the IIR filter shown in Fig. 2.21 and, hence,
find the first *three* terms of its unit sample response.

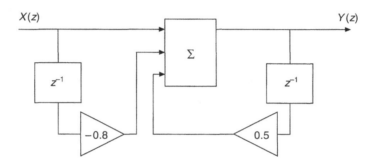

Figure 2.21

2   A simple running average filter generates the average of the current and
previous input samples. Draw the 'z-domain' block diagram of the system.
Hence, derive the transfer function of the filter and use it to find the response
to an input of a discrete unit ramp (first three terms only). ($Y(z) = z^{-1} + 2z^{-2}$
$+ 3z^{-3} + \ldots$ for a unit ramp.)

3   A running average filter averages the current input and the previous *five*
inputs. Show how this system could be realized with a *recursive* filter.

Hint: Find expressions for the output $Y(z)$ and also the output *delayed by one
sample period*, i.e. $Y(z)z^{-1}$ – and then think about it!

4   Figure 2.22 is the z-domain block diagram for a *biquadratic* filter. Show that
its transfer function, $T(z)$, is given by:

$$T(z) = \frac{1 + a_1 z^{-1} + a_2 z^{-2}}{1 - b_1 z^{-1} - b_2 z^{-2}}$$

Hint: Find two simultaneous equations by deriving expressions for $V(z)$ and
$Y(z)$.

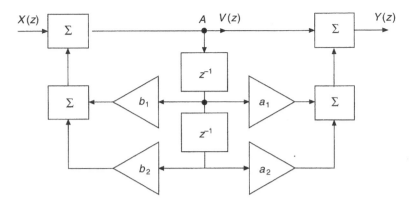

Figure 2.22

2.17 RECAP

- A *non-recursive filter* is one that makes use only of the current input sample and previous input samples. It is also called a finite impulse response filter (FIR) as it generates a finite output sequence in response to a finite input sequence. It has a general transfer function of: $T(z) = a_0 + a_1 z^{-1} + a_2 z^{-2} + a_3 z^{-3} + \ldots$.

- A *recursive filter* is one that has a feedback loop, i.e. it makes use of previous *output* values. It is also called an infinite impulse response filter (IIR) as, *usually*, it generates an infinite output sequence in response to a finite input sequence. If it combines the current input sample with only previous outputs then it has a general transfer function of

$$T(z) = \frac{1}{1 - b_1 z^{-1} - b_2 z^{-2} - b_3 z^{-3} - \ldots}$$

If it also makes use of previous inputs, as is most common, then it has the general transfer function of

$$T(z) = \frac{a_0 + a_1 z^{-1} + a_2 z^{-2} + a_3 z^{-3} + \ldots}{1 - b_1 z^{-1} - b_2 z^{-2} - b_3 z^{-3} - \ldots}$$

- An IIR filter will require fewer coefficients/multipliers than the equivalent FIR filter.

2.18 CHAPTER SUMMARY

In this chapter we have looked at ways of representing discrete signals, both in the time domain and the z-domain. For example, a signal represented by $x[n] = 1, 4, 6, 2$ in the time domain can be expressed as $X(z) = 1 + 4z^{-1} + 6z^{-2} + 2z^{-3}$ in the z-domain. Entering the z-domain allows us to analyse and design digital systems much more easily than working in the time domain.

The z-transform is used to move from the time domain to the z-domain. The z-transform of a sequence $x[n]$ is given by

$$X(z) = \sum_{n=0}^{\infty} x[n]z^{-n}$$

or $x[0]z^0 + x[1]z^{-1} + x[2]z^{-2} + x[3]z^{-3} + \dots$. The 'z^{-n}' operator can be thought of as a tag which indicates a delay of n sample periods.

The transfer function of a system in the z-domain is given by $T(z) = Y(z)/X(z)$, where $X(z)$ and $Y(z)$ are the z-transforms of the input and output sequences respectively. The transfer function is used to find the system response to an input signal.

FIR and IIR filters have been revisited, but this time in the z-domain. FIR filters use the current input sample plus weighted previous inputs to generate the output, while IIR filters use weighted previous outputs as well.

The software package, MATLAB, has been used to analyse discrete systems.

2.19   PROBLEMS

1   An FIR filter has a transfer function of $1 + z^{-1}$. Find the output sequence in response to an input sequence of 1, 2, –2.

2   When the finite sequence 1, –3 is fed into an FIR filter, the output sequence is finite, and equal to 1, –2, –1, –6. Find the transfer function of the filter in the z-domain and also the response of the filter to a discrete unit step (first four terms only).

3   The output sequence, $y[n]$, from a recursive filter, is related to its input sequence, $x[n]$, as: $y[n] = x[n] + 0.5x[n-1] + 0.5y[n-1]$. Draw the block diagram for this filter, *in the z-domain,* and find its transfer function. Use the transfer function to find the filter's response to a unit sample sequence (first four terms only).

4   A filter has a transfer function of $T(z) = (z + 1)(z - 1)/z^2$. Find its response to the finite sequence 1, 0, 2 (first five terms only).

5   Find the transfer function for the processor shown in Fig. 2.23.

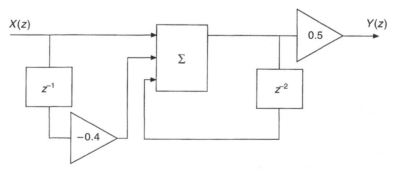

Figure 2.23

6   Use z-transform tables to find the response of a digital filter to a discrete unit step, given that the filter has the transfer function of $0.8/(z - 0.2)$. Only the first four terms are required.

3 The z-plane

3.1 CHAPTER PREVIEW

We start this chapter with a review of the *s-plane,* a concept with which you are probably familiar. We will then move on to its discrete equivalent – the *z-plane*. The *z*-plane has an extremely important role to play in the design and analysis of discrete systems. The significance of *z*-plane *pole–zero diagrams* will be discussed, both in terms of signal shapes and system frequency response. The link between the *s*-plane and *z*-plane will also be examined.

3.2 POLES, ZEROS AND THE s-PLANE

It is very likely that you have already encountered the '*s*-plane'. If so, you will appreciate how useful this concept is when it comes to the analysis and design of analogue systems. In a similar way, the *z-plane* is an invaluable tool when it comes to the analysis and design of *discrete* systems. As it's much easier to explain the significance and the use of the *z*-plane via the *s*-plane, we will begin with a very brief review of the *s*-plane. If you are completely at ease with the Laplace transform, the *s*-plane, and the idea of poles and zeros, then you can probably afford to skip the next few pages. However, if you know little about these, or are very rusty, then you will need to do some extra reading, as it is unlikely that the following review will be sufficient for you. You will find comprehensive explanations in most textbooks dealing with circuit analysis and control systems in particular. Just three of many suitable sources are Bolton (1998), Powell (1995), and Dorf and Bishop (1995).

The *s*-plane is simply an Argand diagram, used to display the values of *s*. Remember that *s* represents the complex frequency, i.e. $s = \sigma + j\omega$, and so *s* values can be naturally shown on an Argand diagram. As is usual for Argand diagrams, the real part of *s*, i.e. σ, is plotted along the horizontal axis and the imaginary part, $j\omega$, along the vertical axis.

As an example, consider a continuous system with a transfer function given by:

$$T(s) = \frac{s + 4}{(s + 2)(s + 3)}$$

This transfer function can be represented on an *s*-plane diagram by displaying its *poles* and *zeros* (Fig. 3.1).

System zeros are the *s*-values that cause the *numerator,* and so $T(s)$ itself, to become zero, while poles are defined as the values of *s* which result in $T(s)$

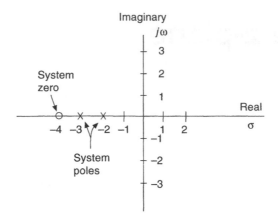

Figure 3.1

becoming infinite, i.e. its *denominator* becoming zero. It follows that this particular transfer function has a single zero at $s = -4$ and two poles at $s = -2$ and -3. On pole–zero diagrams, poles are traditionally represented by crosses and zeros by circles.

In this simple example all values are real and so appear on the real (σ) axis – but what if the transfer function had been more complicated? For example,

$$T(s) = \frac{s^2 + 2s + 5}{(s + 3)(s^2 + 2s + 2)}$$

We will begin by finding the zeros for this transfer function and so must first equate the numerator to zero, i.e. $s^2 + 2s + 5 = 0$. As this quadratic expression factorizes to $(s + 1 + j2)(s + 1 - j2)$, then the system must have complex conjugate zeros at $s = -1 \pm j2$.

To find the poles we need to examine the denominator. Clearly, there must be one pole at $s = -3$ (this is obviously a *real* pole). Also, as $s^2 + 2s + 2$ factorizes to $(s + 1 + j)(s + 1 - j)$, there must be two complex conjugate poles at $s = -1 \pm j$.

The pole–zero diagram for this system is shown in Fig. 3.2.

Pole–zero diagrams of system transfer functions contain a surprising amount of information about the system and are *extremely* useful when it comes to the design and analysis of signal processors. For example, it can be shown that a system with poles in the right-hand half of the *s*-plane is an unstable system. We will examine *system* pole–zero diagrams in much more detail later in the chapter.

3.3 POLE–ZERO DIAGRAMS FOR CONTINUOUS SIGNALS

As signals can also be transformed into the *s*-domain, i.e. expressed as functions of *s*, then they too can be represented by pole–zero diagrams. These 'p–z' diagrams contain information about the signal shapes and so are very useful.

To illustrate this important link between the signal p–z diagram and the signal shape consider a signal, $y(t)$, having a Laplace transform, $Y(s)$, given by:

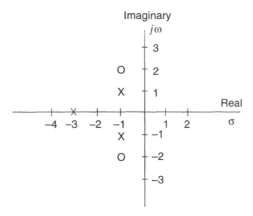

Figure 3.2

$$Y(s) = \frac{3s^2 + 8s + 14}{(s + 2)(s^2 + 2s + 10)}$$

We will first find the time variation of the signal by using the traditional 'inverse Laplace transform' method. However, before the Laplace transform tables can be consulted the expression needs to be broken down into its partial fractions, i.e.:

$$Y(s) = \frac{1}{s + 2} + \frac{2(s + 1)}{s^2 + 2s + 10}$$

(You are strongly advised to derive this expression for yourself.)

By inspection of the Laplace transform tables (Appendix A), you should find that the $1/(s + 2)$ part corresponds to a signal in the time domain of e^{-2t}.

As the other component can be expressed, very conveniently, as $2(s + 1)/[(s + 1)^2 + 3^2]$, it 'inverse Laplace transforms' to $2e^{-t} \cos 3t$. (Again – it's important that you check this for yourself.)

The time variation of the output is therefore given by:

$$y(t) = e^{-2t} + 2e^{-t} \cos 3t \tag{3.1}$$

i.e. an exponentially decaying d.c. signal plus an exponentially decaying sinusoidal (strictly 'cosinusoidal') oscillation.

We now need to look for some link between the signal shape and the pole–zero diagram for Y(s).

As $Y(s)$ factorizes to:

$$Y(s) = \frac{3(s + 1.33 + j1.7)(s + 1.33 - j1.7)}{(s + 2)(s + 1 + j3)(s + 1 - j3)}$$

then $Y(s)$ has zeros at $-1.33 \pm j1.7$ and poles at -2 and $-1 \pm j3$ (Fig. 3.3).

So we have a pole at -2 and a '-2' also appears in the 'e^{-2t}' term of equation (3.1). We also have two complex poles at $-1 \pm j3$. If we look at the $2e^{-t} \cos 3t$ component of the signal, then a '1' occurs in the '$2e^{-t}$' term, i.e. we can think of this as $2e^{-1t}$, and obviously a '3' appears in the 'cos 3t' term. In other words, it's possible that the pole at -2 corresponds to the exponentially decaying d.c. signal

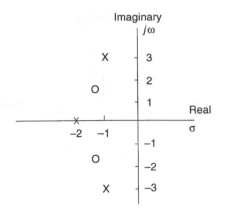

Figure 3.3

of e^{-2t} and the two complex conjugate poles at $-1 \pm j3$ to the $2e^{-t} \cos 3t$ part, i.e. an exponentially decaying cos wave. *This is, in fact, the case.*

All signals represented by poles on the left-hand side of the *s*-plane are decaying signals, the further the poles are away from the imaginary axis, i.e. the more negative the σ-value, the more rapid the decay. If the pole is *on* the real axis then it represents a decaying d.c. signal. Figure 3.4(a) shows the MATLAB p–z diagram

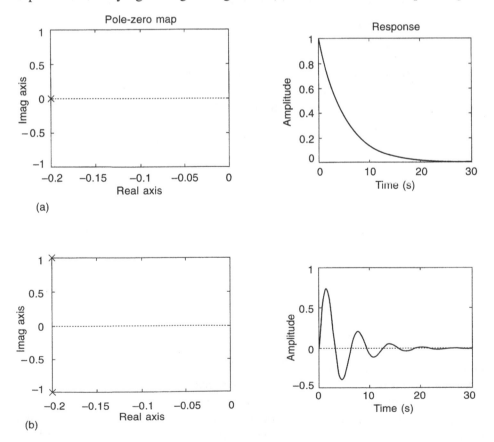

Figure 3.4

and corresponding signal shape for such a signal, with the pole at –0.2. Complex conjugate poles represent a decaying sinusoidal signal – the imaginary coordinates indicating the signal angular frequency, ω (Fig. 3.4(b)). On the other hand (or should it be 'other half'!), a signal having poles on the *right-hand* half of the *s*-plane corresponds to an exponentially *increasing* signal – which is usually extremely undesirable, for obvious reasons. This increasing signal could be sinusoidal (if complex conjugate poles) or d.c. (pole on the real axis). For example, if $Y(s) = 1/(s - 0.2)$, then, from the transform tables, $y(t) = e^{0.2t}$, i.e. an exponentially *increasing* d.c. signal.

If a signal has complex conjugate poles *on* the imaginary axis, at $\pm j\omega$, then this corresponds to a non-decaying sinusoidal signal of angular frequency, ω. For example, if the poles are at $\pm j0.2$, i.e.

$$Y(s) = \frac{1}{(s + 0.2j)(s - 0.2j)} = \frac{1}{(s^2 + 0.04)}$$

then inspection of the Laplace tables gives us $y(t) = 5 \sin 0.2t$.

For the special case of a single pole at the origin, i.e. $Y(s) = k/s$, this represents a step of height k.

Why have we concentrated on the poles and ignored the zeros? This is because the zeros do not affect the general shape of the individual signal components corresponding to poles, *but only their weightings* – the closer a zero is to a pole, the smaller the signal corresponding to that pole.

3.4   SELF-ASSESSMENT TEST

1 A signal has a Laplace transform given by

$$Y(s) = \frac{2(s + 2)}{s(s^2 + 4s + 3)}$$

Draw the p–z diagram and hence sketch the rough shapes of the individual components that make up the signal (there will be three of these). Now use partial fractions and the Laplace transform tables to find the time variation of the signal.

2 A signal has the Laplace transform of $6/(s^2 + 4)$. Use the pole positions to predict the rough shape of the signal and then use Laplace transform tables to find the exact shape.

3.5   RECAP

When the Laplace transform of a signal is represented by a p–z diagram:

- The negative *real component, σ,* of a pole position, corresponds to an exponential decay – the further the pole is from the imaginary axis, the more rapid the decay of the signal.

- The *imaginary components,* $\pm j\omega$, of conjugate pole positions, correspond to a sinusoidal oscillation of angular frequency ω, while a pole *on* the real axis corresponds to a d.c. signal.

- If a pair of conjugate poles lies *on* the imaginary axis, i.e. have no real part, then they represent a non-decaying oscillation. The special case of a pole at the origin corresponds to a step function, i.e. a non-decaying d.c. signal.

- If the signal has a pole (or complex conjugate poles) on the right-hand side of the pole–zero diagram, i.e. a positive σ-value, then this corresponds to a signal which is *increasing* exponentially, i.e. *bad news*! (In a similar way, if the *transfer function* of a system has poles on the right-hand side, then this is an unstable system and must be redesigned.)

- The zeros do not affect the shapes of the components making up the signals *but only their weightings* – the nearer a zero is to a pole the less the contribution made by that pole to the signal.

3.6   FROM THE *s*-PLANE TO THE *z*-PLANE

We have seen that we can represent *continuous* signals and systems by their *s*-plane, pole–zero diagrams. Further, if a signal has a pole on the right-hand side of the *s*-plane, then this represents an exponentially *increasing* signal. Similarly, if a transfer function has a pole in the right-hand side, then this indicates an unstable system (Fig. 3.5); again, something best avoided!

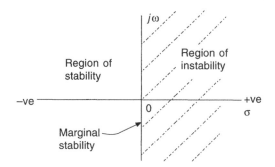

Figure 3.5

In a similar way, we can represent *discrete* signals and systems by their p–z diagrams *in the z-plane*. For example, if a discrete signal has a *z*-transform, $X(z)$, given by:

$$X(z) = \frac{z - 0.2}{z^2 + 0.6z + 0.13}$$

then there will be a zero at 0.2 and, as the denominator factorizes to $(z + 0.3 \pm j0.2)$, complex conjugate poles at $-0.3 \pm j0.2$. The p–z diagram is shown in Fig. 3.6.

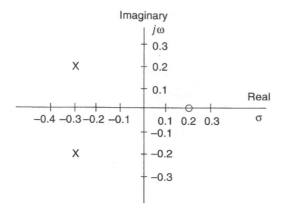

Figure 3.6

We know that the locations of signal poles and zeros in the *s*-plane give us a good idea of the signal shape – but how do we interpret these positions in the *z*-plane? Also, which regions in the *z*-plane correspond to stability and instability?

To answer these questions we need to start with the link between the *s*-plane and the *z*-plane. Although we have used the z^{-n} operator, and appreciate that it indicates a delay of *n* sampling periods, no mention has yet been made of any link between *z* and *s*. In fact there is a relationship between them and it is an extremely important one. This is given by:

$$z = e^{sT}$$

where *T* is the sampling period. (Just accept this relationship for now – it will be explained later in the chapter.)

Using this relationship we will first consider the issue of stability.

3.7 STABILITY AND THE *z*-PLANE

We know that the imaginary axis in the *s*-plane is the boundary between the areas of stability and instability and that this boundary is defined by $s = j\omega$. From $z = e^{sT}$, this boundary must therefore map on to the *z*-plane as:

$$z = e^{j\omega T}$$

In other words, this equation defines the boundary between the stable and unstable regions in the *z*-plane.

O.K. – but what does this boundary look like? In fact, the equation $z = e^{j\omega T}$ defines a circle, centred on the origin, and with a radius of 1.

This is far from obvious and, to prove it, we need to use *Euler's identity*. This tells us that:

$$e^{j\theta} = \cos\theta + j\sin\theta$$

It follows that we can rewrite $z = e^{j\omega T}$ as:

$z = \cos \omega T + j \sin \omega T$

Figure 3.7 shows the point representing this z-value, plotted on the Argand diagram.

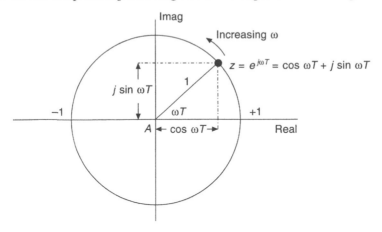

Figure 3.7

From the theory of complex numbers, or just from Fig. 3.7, the length of Az must be given by $\sqrt{\cos^2 \omega T + \sin^2 \omega T}$, which is equal to 1, i.e.

$|z| = 1$

Also, the tan of the angle made by Az with the real, positive axis is given by

$$\frac{\text{imaginary part}}{\text{real part}} = \frac{\sin \omega T}{\cos \omega T} = \tan \omega T$$

and so this angle must be ωT.

From Fig. 3.7 it should be clear that these magnitude and angle values do indeed define a unit circle in the z-plane, with its centre at the origin. This is because the length of the line Az, from the origin to the z-value, is always 1, and it makes an angle of ωT with the positive real axis. Therefore, as the signal frequency increases, i.e. as ω gets larger, z must mark out a circle of radius 1.

So we now know that the imaginary axis in the s-plane transforms to this 'unit circle' in the z-plane, i.e. *this circle is the boundary between the stable and unstable regions in the z-plane.* This leaves just one obvious question to be answered: 'Is the stable region inside or outside the circle?' The easiest way to decide on this is to map an 'unstable' pole in the s-plane, onto the z-plane, and see where it lies. *Any* positive s-value corresponds to a point on the unstable right-hand half of the s-plane. As $s = 0$ corresponds to a z-value of 1 ($e^0 = 1$), any *positive* s-value must therefore transform to a z-value which is *greater* than 1 – i.e. *to a point outside the unit circle.*

It follows that, for the z-plane, it is the area outside the unit circle which is the region of instability and the area enclosed by the circle the region of stability – the unit circle itself corresponds to marginal stability (as does the imaginary axis in the s-plane).

These important 's-to-z' mappings are summarized in Fig. 3.8.

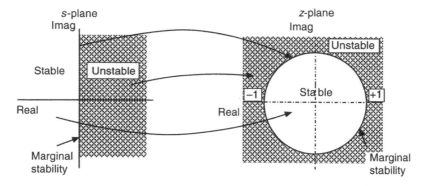

Figure 3.8

3.8 DISCRETE SIGNALS AND THE z-PLANE

We now know that the area outside the unit circle in the z-plane corresponds to the right-hand half of the s-plane. In other words, if the z-domain transfer function of a discrete system has a pole outside the unit circle in the z-plane, then this is an unstable system. It also means that if the z-transform of a discrete signal has a pole in this area, then the signal amplitude increases with time.

But how, exactly, is the shape of the discrete signal related to its z-plane, pole–zero diagram? To answer this question, we need to start with an s-plane pole of a continuous signal. We know that the position of an s-plane pole can be expressed, completely generally, as $s = \sigma + j\omega$.

So, from $z = e^{sT}$, an s-plane pole will map to the z-plane as:

$$z = e^{(\sigma + j\omega)T} = e^{\sigma T} e^{j\omega T}$$

We now need to look, in more detail, at the two exponential expressions, $e^{\sigma T}$ and $e^{j\omega T}$, that make up z.

From earlier work we know that we can replace the $e^{j\omega T}$ part by cos ωT + j sin ωT (Euler's identity), i.e. it indicates a point that is unit distance from the origin. Further, the line joining this point to the origin makes an angle of ωT with the positive real axis.

The '$e^{\sigma T}$' part is real, and so just corresponds to a real number. For example, if $T = 0.1$ s and $\sigma = -2$, then $e^{\sigma T} \approx 0.82$. This part therefore acts as a multiplier and must fix the distance from the pole to the origin (Fig. 3.9). The line joining the pole to the origin is usually referred to as the *pole vector*. We know that, for *continuous signals,* σ indicates how rapidly the signal decays or increases. It therefore seems reasonable that the more negative is σ, i.e. the shorter the distance from the pole to the origin, the more rapidly the discrete signal decays.

To summarize, it is the *angle* made by the pole vector with the positive real axis that represents the signal frequency, while the *length* of the vector tells us how quickly the signal is decaying or increasing – the shorter the vector the more rapid the decay rate.

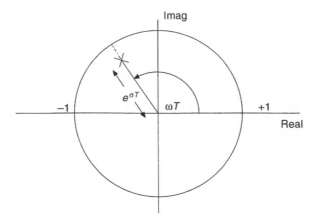

Figure 3.9

To make this clearer, let's imagine that we have a discrete signal that has a z-transform of $1/(z - 0.5)$. The simple p–z diagram is shown in Fig. 3.10.

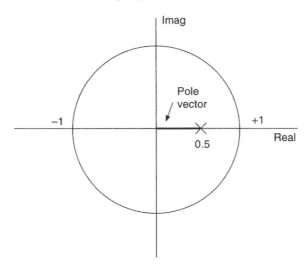

Figure 3.10

As the pole vector from the pole to the origin makes an angle of zero with the positive real axis, then this must represent a discrete signal of frequency 0 Hz, i.e. a d.c. signal. Also, as the pole vector has a length of 0.5, *then this indicates that the signal should fall to a half of its previous value every sampling period.* Figure 3.11 is the MATLAB plot for this signal – encouragingly it agrees with our prediction.

To confirm this plot we can simply carry out the division of $1/(z - 0.5)$. This gives $z^{-1} + 0.5z^{-2} + 0.25z^{-3} + 0.125z^{-4} + \ldots$, which agrees with the MATLAB plot.

Notice that, although it has the expected rate of decay, the signal is delayed by one sample period, i.e. it starts at $t = T$ and not $t = 0$. This is not immediately obvious from the p–z diagram, but *is* predicted by the long division.

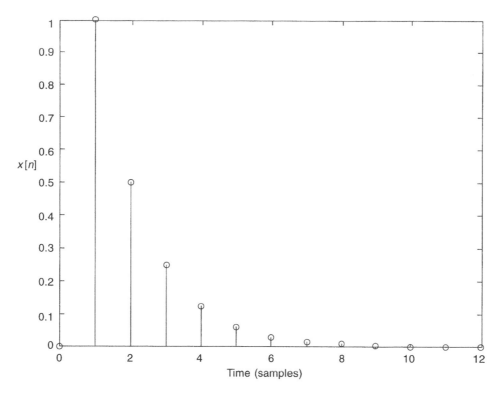

Figure 3.11

This was a fairly simple example, having real roots, but what if the signal z-transform had been something more complicated, such as $1/(z^2 - z + 0.5)$?

As the denominator has the complex conjugate roots of $0.5 \pm j0.5$, then the p–z diagram is as shown in Fig. 3.12. The pole vector angle is 45° (and so $\omega T = 45°$ or $\pi/4$ rad. (N.B. It is the angle made by the pole vector in the top half of the z-plane which is important, i.e. 45° rather than the 315° or –45° in the bottom half.)

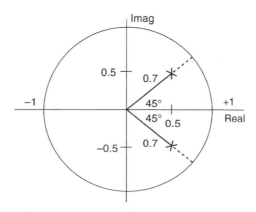

Figure 3.12

If we use a sampling period of 0.1s then this gives:

$0.1\omega = \pi/4$

or $\omega = 2.5\pi$ rad/s, which is equivalent to a frequency of 1.25 Hz, as $\omega = 2\pi f$.

The pole vector has a length of 0.7, and so we would expect the discrete signal to be a decaying sinusoid of frequency 1.25 Hz, the sinusoid decaying by a factor of 0.7 every 0.1 s.

The MATLAB plot, Fig. 3.13, certainly confirms the signal frequency, i.e. it has a period of 0.8 s and so a frequency of 1.25 Hz. It also agrees with the expected rate of decay – although the latter is not quite so easy to identify. Again note the delay of one sample period.

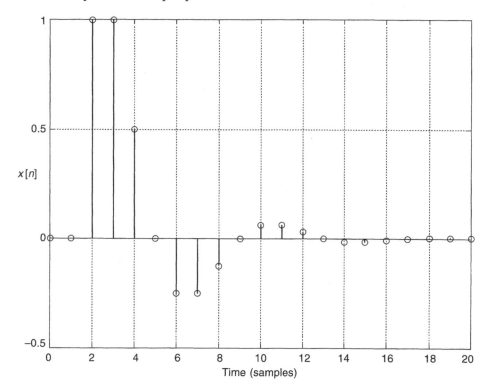

Figure 3.13

N.B.   If the pole-pair (or single pole if d.c.) lies *on* the unit circle, then this corresponds to an *undamped* signal, i.e. one with a constant amplitude, while poles *outside* the unit circle correspond to a signal of *increasing* amplitude. For example, if the length of the pole vector is 1.2, then the signal increases by a factor of 1.2 for each sample period. This signal could be a d.c. signal or sinusoidal. Obviously, if the signal is sinusoidal then this increase is superimposed on to the normal sine variation, as is a decay for poles *inside* the unit circle.

3.9   ZEROS

You will remember that signal zeros in the *s*-plane do not change the overall shape of the signal but they do affect the *magnitude* – the nearer a zero is to a pole, the less the size of the corresponding signal component. Zeros have a similar effect in the *z*-plane *but here they also affect signal time delays.* Earlier we looked at a signal with a *z*-transform of $1/(z - 0.5)$. This signal is a decaying d.c. waveform with a delay of one sample period (Fig. 3.11). What if the *z*-transform had been $z/(z - 0.5)$? The only difference with this signal is that it has a zero at the origin. The corresponding signal is shown in Fig. 3.14. Note that

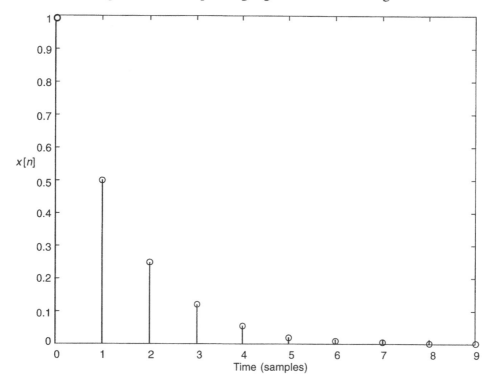

Figure 3.14

there is now no time delay, i.e. the effect of the zero at the origin is to eliminate the time delay of one sampling period. This can easily be confirmed by dividing *z* by $(z - 0.5)$ which gives $1 + 0.5z^{-1} + 0.25z^{-2} + \ldots$ rather than $z^{-1} + 0.5z^{-2} + 0.25z^{-3} + \ldots$. So the very existence of the zero will get rid of this time delay. Further, if the zero had not been at the origin, its closeness to the pole would also have affected the signal size. For example, Fig. 3.15 shows the signal corresponding to a *z*-transform of $(z - 0.2)/(z - 0.5)$, i.e. the presence of this zero results in a greater rate of decay between the first and second samples – the closer the zero to the pole, the greater will be the reduction. Again, there is no delay.

Question
What would happen to the rate of decay if the zero were in the other half of the z-plane to the pole, for example, at –0.2?

Answer
If you carry out a long division you should find that the signal has a z-transform of $1 + 0.7z^{-1} + 0.35z^{-2}$, i.e. surprisingly perhaps, a zero in the other half of the z-plane to the pole actually reduces the initial rate of decay.

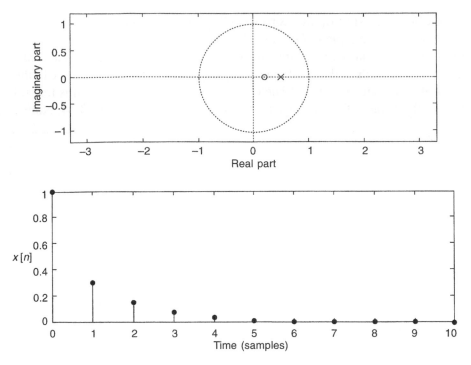

Figure 3.15

N.B.   MATLAB p–z diagrams are obtained using the 'z-plane' function. For example:

```
num=[1 -0.2];
den=[1 -0.5];
zplane(num,den)
```

was used to produce the p–z plot of Fig. 3.15. Displaying more than one plot at the same time, i.e. the p–z plot and the corresponding signal, can be achieved by using the MATLAB 'subplot' function. Refer to the user manual for further details.

3.10   THE NYQUIST FREQUENCY

Consider a discrete signal represented by a pole on the *negative real axis*. For example, let the signal z-transform be given by $X(z) = 1/(z + 0.5)$. This signal will have a z-plane pole at -0.5 and so the pole vector angle is 180° or π radians (Fig. 3.16).

Let's assume a sampling frequency of 10 Hz, i.e. $T = 0.1$ s. As the phase angle $(\theta) = \omega T$, then $\pi = 0.1\omega$ and so this pole indicates a signal with angular frequency of 10π, or $f = 5$ Hz.

Note that this frequency is half the sampling frequency of 10 Hz, i.e. it is the *Nyquist* frequency – the maximum signal frequency if aliasing is to be avoided. This is not just a coincidence but will always be the case, no matter what the sampling frequency.

A signal represented by a pole on the negative real axis has a frequency equal to the Nyquist frequency.

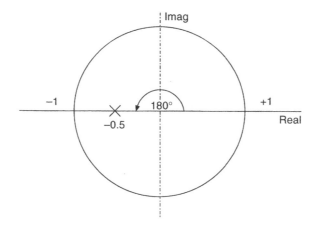

Figure 3.16

3.11   SELF-ASSESSMENT TEST

1 Sketch the signal represented by the p–z diagram shown in Fig. 3.17, *by inspection of the diagram only*. Then check your prediction by carrying out a suitable polynomial division.

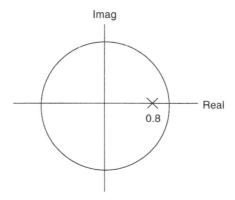

Figure 3.17

2 Each of the poles/pole-pairs shown in Fig. 3.18 represents a separate discrete signal. Given that the sampling frequency is 120 Hz, state the frequencies of each of the signals.

3 Map the poles/pole-pairs shown on the *s*-plane diagram of Fig. 3.19, onto *a z*-plane diagram, assuming a sampling frequency of 2 Hz. (N.B $z = e^{sT}$.)

3.12   THE RELATIONSHIP BETWEEN THE LAPLACE AND *z*-TRANSFORM

Earlier, I promised to show where the vital relationship $z = e^{sT}$ comes from – well this is it!

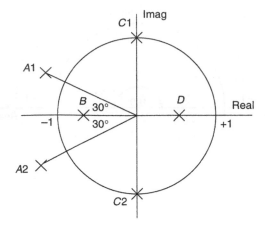

Figure 3.18

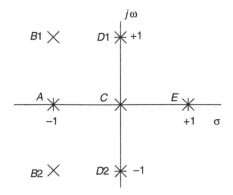

Figure 3.19

To be able to do this, we first need to go back to the definition of the Laplace transform, $X(s)$, for a *continuous* signal, $x(t)$, i.e.

$$X(s) = \int_0^\infty x(t)e^{-st}\,\mathrm{d}t$$

The *discrete* version of this signal, i.e. the signal sampled every T seconds, will only exist at time $= 0, T, 2T, 3T$, etc. It follows that we *first need to replace t with nT, where n is 0, 1, 2, 3 . . . .*

Also, the integral sign, which indicates the area under the $x(t)$ plot, has no meaning when dealing with discrete signals, as they consist of infinitesimally thin pulses. The equivalent process is the *addition* of the pulse magnitudes and so *we must replace the integral symbol with the summation symbol.*

The Laplace transform then becomes:

$$X^*(s) = \sum_{n=0}^\infty x[n]e^{-snT}$$

where X^* indicates the *discrete* version of the Laplace transform.

If we now define the operator z as:

$$z = e^{sT}$$

then the transform becomes:

$$X(z) = \sum_{n=0}^{\infty} x[n]z^{-n} = x[0] + x[1]z^{-1} + x[2]z^{-2} + \ldots$$

which you should recognize as the definition of the z-transform.

In other words, it is this particular relationship of $z = e^{sT}$ which causes the z-transform to emerge from the Laplace transform.

3.13 RECAP

- The z-plane plays a similar role in the design and analysis of discrete systems to that of the s-plane in the design and analysis of continuous systems.

- Points on the s-plane map on to the z-plane as $z = e^{sT}$.

- Signals represented by poles within, on and outside the z-plane unit circle are decaying, non-decaying and increasing in amplitude respectively.

- The frequency of a discrete signal is obtained from the angle made by the pole vector with the positive real axis, while the vector magnitude indicates the rate of decay – the shorter the vector the greater the rate of decay. A selection of p–z diagrams along with the corresponding signals is shown in Fig. 3.20.

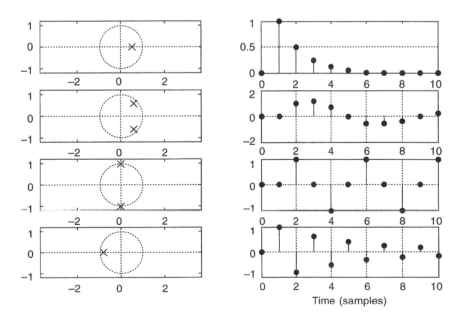

Figure 3.20

3.14   THE FREQUENCY RESPONSE OF CONTINUOUS SYSTEMS

We have come this far through a book on digital signal processing – which is largely to do with digital filters – and there has been no mention made of *frequency response*, even though the concept of frequency response is pretty fundamental to filter analysis and design! I assure you that this has not been an oversight on my part. It is just that we are only now in a position to make much sense of the frequency response of *discrete* systems.

Before we look at this very important topic it's best to take a step backwards, and first remind ourselves of how we can find the frequency response of *continuous* systems. We will also look at how we can get this information from *s*-plane p–z diagrams.

The frequency response of a system is the way in which the gain of the system, and also the phase difference between the output and input signals, depend on the signal frequency. The test signal is always a non-decaying, pure sinusoidal waveform.

Suppose we have an analogue system with a transfer function given by $T(s) = (s + 1)/(s + 2)$, and we need to find its frequency response. We know that s is the complex frequency of a signal, i.e. $s = \sigma + j\omega$, where σ indicates the rate of decay of the signal and ω is its angular frequency. However, as we are only interested in the response to a test signal which is a *non-decaying* sine wave, then $\sigma = 0$ and so we can replace s with just $j\omega$.

We now have

$$T(j\omega) = \frac{j\omega + 1}{j\omega + 2}$$

To find the *gain* of the system we need to find the magnitude of this complex function, i.e.

$$| T(j\omega) | = \frac{\sqrt{\omega^2 + 1^2}}{\sqrt{\omega^2 + 2^2}} \tag{3.2}$$

To find the *phase relationship* between output and input we must get the phase angle of $T(j\omega)$. From complex number theory, this phase angle is given by:

$$\angle T(j\omega) = \tan^{-1} \frac{\omega}{1} - \tan^{-1} \frac{\omega}{2} \tag{3.3}$$

For example, let's imagine that we are interested in the gain at an angular frequency of 1 rad /s. Substituting $\omega = 1$ into equations (3.2) and (3.3) gives a gain of $\sqrt{2/5}$ or 0.63 and a phase angle, $\angle T(j\omega)$, of $\tan^{-1} 1 - \tan^{-1} \frac{1}{2} = 45° - 26.6° = 18.4°$, i.e.

$$T(j\omega) = 0.63\angle 18.4°$$

You are probably very familiar with what we have just done, as this is the normal method used to find the frequency response of a system. However, there is an alternative way of doing this which makes use of the p–z diagram for the system. This method can often be more convenient, especially when we want to design or analyse filters.

To demonstrate this approach we will stick with the same system, i.e. $T(s) = (s + 1)/(s + 2)$. First its p–z diagram must be drawn (Fig. 3.21). This is a particularly simple one, having a zero at –1 and a pole at –2.

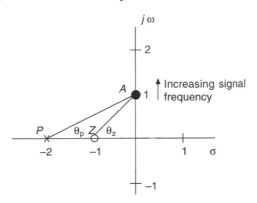

Figure 3.21

The gain and phase angle at 1 rad/s will again be found – we can then check against the values obtained earlier, using the 'traditional' method. To do this, a point A needs to be placed at position $(0 + j)$.

N.B. This point represents a non-decaying, sinusoidal signal of angular frequency 1 rad/s, $(s = j\omega = j)$.

To get the gain of the system, all we need to do is find the distances from A to the zero and to the pole and then divide these two distances. The zero distance is $\sqrt{1^2 + 1^2}$ and the pole distance is $\sqrt{1^2 + 2^2}$ and so the gain is $\sqrt{2/5} = 0.63$, which is what we got using the 'normal' method. This isn't too surprising as, for any ω value, the length of the zero vector from the zero to A is exactly the same as the magnitude of the numerator of equation (3.2), i.e. $\sqrt{\omega^2 + 1^2}$, while the length of the pole vector is the same as the magnitude of the denominator, $\sqrt{\omega^2 + 2^2}$.

To find the phase angle we just measure the angles between the positive real axis and both the zero vector and the pole vector, and then subtract these two angles. As the zero angle, θ_z, is $\tan^{-1}1$ (45°), and the pole angle, θ_p, is $\tan^{-1}1/2$ (26.6°), then this gives a phase angle of 18.4° – exactly the same as we derived earlier using the other method.

This graphical method gives a general expression for the phase angle of $\tan^{-1}(\omega/1) - \tan^{-1}(\omega/2)$, which, once again, is exactly the same as we obtained using the more traditional approach (equation (3.3)).

This graphical approach is very useful in that we can get a good idea of the frequency response just by inspection of the positions of the poles and zeros. For example, if a pole lies very near to the imaginary axis at $\omega = 3$ rad/s say, then we would expect a large gain at this frequency, while if it had been a zero, then the gain will be low. With experience it is possible to place poles and zeros so as to design a filter with the required response – this is a particularly useful technique when designing *digital* filters.

Example 3.1

Find the gain and phase angle for the previous system, i.e. $T(s) = (s + 1)/(s + 2)$, for angular frequencies of 0 rad/s (d.c.), 2 rad/s and also for very large frequencies.

Solution

- $\omega = 0$ rad/s: this frequency (d.c.) is represented by point A in Fig. 3.22.

  $\text{Gain} = \dfrac{AZ}{AP} = \dfrac{1}{2} = 0.5$ and phase angle $= 0 - 0 = 0°$

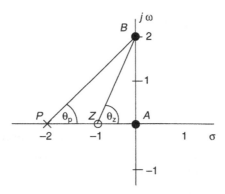

Figure 3.22

- $\omega = 2$ rad/s: this frequency is represented by point B:

  $$\text{Gain} = \frac{BZ}{BP} = \frac{\sqrt{1^2 + 2^2}}{\sqrt{2^2 + 2^2}} = \sqrt{\frac{5}{8}} \approx 0.8$$

  Phase angle $= \theta_z - \theta_p = \tan^{-1} 2 - \tan^{-1} 1 = 18.4°$.

- *For large frequencies:* here, both zero and pole distances will be very large and approximately the same size. It follows that the gain must become very close to unity. Similarly, both zero and pole angles will approach 90° for sufficiently large frequencies and so the phase angle will be approximately zero.

  Armed with this information, it should be possible to make a reasonable stab at sketching the frequency response. Figure 3.23 shows precise MATLAB plots, produced using the 'bode' function, i.e.:

```
num=[1  1];
den=[1  2];
bode(num,den)
```

Remember that a Bode plot displays both the gain and the phase angle against $\log \omega$ (or $\log f$), the gain being in dB (20 log gain). So we are looking for gains

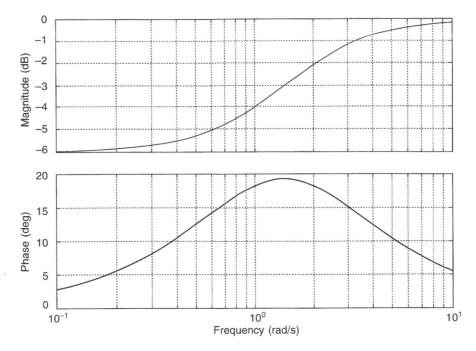

Figure 3.23

of 20 log 0.5, 20 log 0.63, 20 log 0.8 and 20 log 1 at angular frequencies of 0, 1 and 2 rad/s and very large frequencies respectively, i.e. approximately −6 dB, −4 dB, −1.94 dB and 0 dB. Our predicted gain and phase values agree closely with the Bode plot values. Note that it is impossible to display 0 rad/s on a log scale. However, our gain has clearly settled at −6 dB by 0.1 rad/s and the phase angle appears to be approaching 0.

N.B.   In this particular example we only had one pole and one zero. For the more general case, where there are several zeros and poles:

$$\text{Gain} = k \, \frac{\Pi \text{ zero distances}}{\Pi \text{ pole distances}}$$

$$\text{Phase angle} = \Sigma \text{ zero angles} - \Sigma \text{ pole angles}$$

where k is the 'pure' gain in the system. For example, if the transfer function of the system had been $3(s + 1)/(s + 2)$, then $k = 3$.
'Π' is the shorthand symbol commonly used to represent 'product of', in the same way as 'Σ' is used to indicate the sum.
If there is no zero, e.g. $T(s) = 1/(s + 2)$, then a '1' must be used for the zero distance.

3.15   SELF-ASSESSMENT TEST

A continuous system has a transfer function given by $T(s) = 4(s + 1)(s + 2)/(s^2 + 2s + 2)$. Plot its p–z diagram and use this diagram to find the response, both gain and phase, at 0, 1 and 2 rad/s and also for very large frequencies. Also, sketch the complete frequency response.

3.16   THE FREQUENCY RESPONSE OF DISCRETE SYSTEMS

As with continuous systems, the frequency response of a *discrete system* can also be found from its pole–zero diagram. However, as it is the unit circle in the *z*-plane that corresponds to the imaginary axis in the *s*-plane, *here the 'frequency point' moves around the unit circle in an anticlockwise direction as the signal frequency increases.* The point starts at $z = +1$ (d.c.), and ends at $z = -1$ (the Nyquist frequency).

For example, say we have a discrete system with a transfer function given by $T(z) = (z + 1)/(z - 0.5)$, and we need to find the response at d.c. (0 Hz), 1 Hz and 2 Hz, given that the sampling frequency is 8 Hz.

The p–z diagram is shown in Fig. 3.24. As the sampling frequency is 8 Hz, the Nyquist frequency is 4 Hz, and so the three frequencies of 0, 1 Hz and 2 Hz must correspond to points *A, B* and *C* respectively.

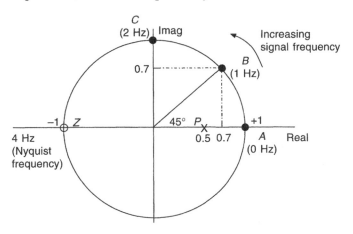

Figure 3.24

- Frequency = 0 Hz:

$$\text{Gain} = k \frac{\Pi \text{ zero distances}}{\Pi \text{ pole distances}}$$

∴ Gain = *AZ*/*AP* ($k = 1$ here)

∴ Gain = 2/0.5 = 4

Phase angle = Σ zero angles − Σ pole angles

∴ Phase angle = $0 - 0 = 0°$

- Frequency = 1 Hz: The pole and zero distances are more difficult to calculate here – using a scale diagram and measuring the distances is an alternative approach. However, whichever method is used, the zero distance, *BZ*, ≈ 1.84 and the pole distance, *BP*, ≈ 0.73. Check these for yourself.

∴ Gain = 1.84/0.73 ≈ 2.5

Also the zero angle $\approx \tan^{-1}(0.7/1.7) = 22.3°$ and the pole angle $\approx \tan^{-1}(0.7/0.2)$ $= 74.1°$.

$\therefore$ Phase angle $\approx 22.3° - 74.1° = -51.8°$

- Frequency $= 2$ Hz: Here the zero distance is $\sqrt{2}$ and the pole distance is $\sqrt{1^2 + 0.5^2}$ (Fig. 3.25).

$\therefore$ Gain $= \dfrac{\sqrt{2}}{\sqrt{1.25}} = 1.27$

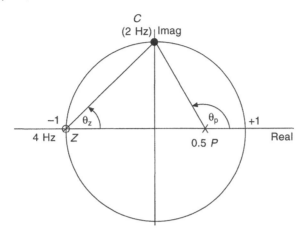

Figure 3.25

The zero angle is $45°$, and the pole angle is $\{180° - \tan^{-1}(1/0.5) = 116.5°\}$.

$\therefore$ Phase angle $= 45° - 116.5° = -71.5°$

N.B.   Remember that the phase angle is always measured between the horizontal, *pointing to the right,* and the vector, measured from the horizontal to the vector moving in an *anticlockwise* direction.

Figure 3.26 shows the MATLAB frequency response plots for the system, drawn using the 'freqz' function. Note that with the 'freqz' function the frequency axis is linear and also 'normalized', i.e. the frequency of '1' corresponds to the Nyquist frequency.

Our gain values agree very well with the magnitude plot of Fig. 3.26, i.e. at 0 Hz, 1 Hz ($0.25f_N$) and 2 Hz ($0.5f_N$), we get gains of 4, 2.5 and 1.27 respectively, or 12 dB, 8 dB and 2 dB. Our corresponding phase angles of 0, $-51.8°$, and $-71.5°$ are also in good agreement.

The program used to produce these plots is:

```
num=[1  1];
den=[1  -0.5];
freqz(num,den)
```

N.B.   Because the frequency axes of the plots generated by the 'freqz' function are linear rather than logarithmic, these plots are not strictly Bode plots. There *is* a MATLAB Bode plotting function for discrete systems, which could have been used. This is the 'dbode' function. This alternative display (Fig. 3.27) was produced using:

```
num=[1 1];
den=[1 -0.5];
T=0.125; %Sampling period
dbode(num,den,T)
```

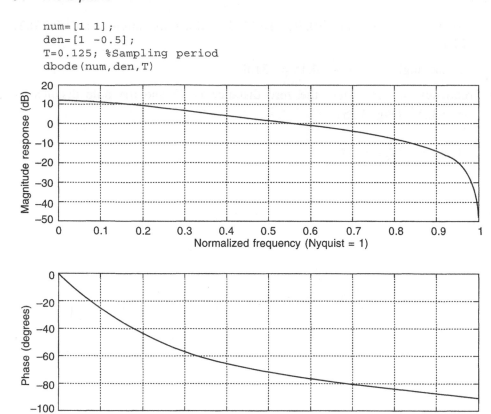

Figure 3.26

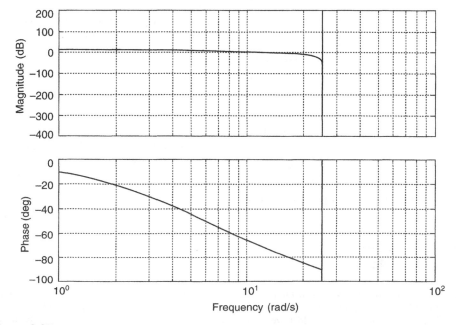

Figure 3.27

However, this function is not ideal as the gain is shown as becoming very low at the Nyquist angular frequency of approximately 25 rad/s. Because of this huge magnitude range, values of interest are extremely compressed. This problem can be overcome to some extent but, generally, the 'freqz' function is much more convenient.

Notice that the frequency response is only displayed up to the Nyquist frequency on both of the freqz and the dbode plots. The response does not stop at this frequency, of course, but continues from f_N to f_s, i.e. as the frequency point proceeds around the unit circle. Due to the symmetry of the p–z diagram about the real axis, this portion of the magnitude response is a mirror image of that up to f_N. The frequency response then repeats itself as the frequency point makes a second revolution of the unit circle, corresponding to frequencies from f_s to $2f_s$ – and so on. However, you need to be clear that this is the frequency response of the *digital processor* at the heart of the DSP system. The output from the *complete* system will only contain signal components with frequencies up to f_N. This is because the reconstruction filter at the output of the DSP system is a lowpass filter, with a cut-off frequency of f_N (see 'The reconstruction filter', Section 1.4). For this reason we have little interest in the frequency response above the Nyquist frequency.

Alternative calculation of frequency response

As an alternative to using the p–z diagram, the frequency response of a *continuous* system can be calculated directly from its transfer function by replacing s with $j\omega$. We can also calculate the frequency response of a discrete system directly from *its* transfer function but, as usual, it's a bit more complicated.

As an example, let's take the system considered earlier in this section, i.e. the one with the transfer function $(z + 1)/(z - 0.5)$ and with $T = 0.125$ s.

We first need to move from the z-domain to the s-domain using the link provided by $z = e^{sT}$. However, as we are interested in the *frequency response,* then $s = j\omega$, and so we need to replace z with $e^{j\omega T}$.

Substituting this z-value into the transfer function gives

$$T(j\omega) = \frac{e^{j\omega T} + 1}{e^{j\omega T} - 0.5}$$

This expression doesn't look too hopeful! However, we can improve things by applying Euler's identity, i.e. $e^{j\theta} = \cos\theta + j\sin\theta$. This gives

$$T(j\omega) = \frac{\cos\omega T + j\sin\omega T + 1}{\cos\omega T + j\sin\omega T - 0.5}$$

which allows the gain and phase angle to be obtained from $\omega = 0$ to $\omega = \pi/T$ (the Nyquist angular frequency).

For example, if we want the response at 1 Hz (2π rad/s) say, then as $T = 0.125$ s:

$$T(j\omega) = \frac{\cos 0.785 + j\sin 0.785 + 1}{\cos 0.785 + j\sin 0.785 - 0.5} = \frac{1.71 + j0.71}{0.21 + j0.71}$$

which has a more familiar format.

$$\therefore \text{Gain} = \frac{\sqrt{1.71^2 + 0.71^2}}{\sqrt{0.21^2 + 0.71^2}} = 2.5$$

and

$$\text{Phase angle} = \tan^{-1} \frac{0.71}{1.71} - \tan^{-1} \frac{0.71}{0.21} \approx -51°$$

N.B.   Before moving on, it would be worth using this method to also check the frequency response at 0 and 2 Hz for yourself.

Example 3.2

A digital filter has a transfer function of $3(z - 0.5)(z + 1)/(z^2 + 0.25)$. Find the frequency response at 0 Hz, and 2.5 Hz, (a) by using the p–z diagram and (b) directly from the transfer function. The sampling frequency is 10 Hz.

Solution

(a) *From the p–z diagram.*
- This system has zeros at 0.5 and –1 and poles at $\pm j0.5$. The p–z diagram is shown in Fig. 3.28.

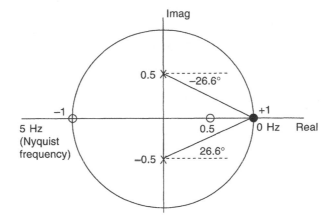

Figure 3.28

- $f = 0$ Hz:

  $$\text{Magnitude} = k \frac{\Pi \text{ zero distances}}{\Pi \text{ pole distances}}$$

  $$\therefore \text{Magnitude} = \frac{3 \times 0.5 \times 2}{\sqrt{1.25} \ \sqrt{1.25}} = 2.4$$

  Phase angle = Σ zero angles – Σ pole angles

  $\therefore$ Phase angle = $0 + 0 - \tan^{-1} 0.5 - (360° - \tan^{-1} 0.5) = 360°$ or $0°$

- $f = 2.5$ Hz: from Fig. 3.29,

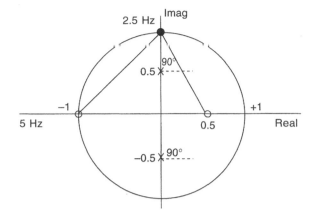

Figure 3.29

Magnitude $\dfrac{3 \times \sqrt{1.25} \times \sqrt{2}}{0.5 \times 1.5} = 6.3$

Phase angle $= \tan^{-1} 1 + (180° - \tan^{-1} 2) - 90° - 90°$

$$= 45° + (180° - 63.4°) - 180° = -18.4°$$

(b) *Directly from the transfer function.*

- $f = 0$ Hz: from $z = e^{j\omega T}$, with $\omega = 0$, we get $z = 1$. (We didn't really need to use $z = e^{j\omega T}$ to find that $z = 1$ corresponds to 0 Hz.)

Substituting $z = 1$ into the transfer function:

$$T(z) = \frac{3 \times 0.5 \times 2}{1.25}$$

This is a real function and so gain $= 2.4$ and phase angle $= 0$ (as the imaginary component is zero).

- $f = 2.5$ Hz: $\omega = 2\pi f = 5\pi$ rad/s. We could use $z = e^{j\omega T}$ to find the z-value corresponding to this particular frequency, but this is another frequency where we don't have to bother as it's obvious that, as 2.5 Hz is a half of the Nyquist frequency, $z = j$.

Substituting $z = j$ into the transfer function:

$$T(j\omega) = \frac{3(j - 0.5)(j + 1)}{(-1 + 0.25)} \approx \frac{3(1.12\angle116°)(1.4\angle45°)}{-0.75} = \frac{4.7\angle161°}{0.75\angle180°}$$

$$= 6.3\angle-19°$$

(N.B. The denominator of -0.75 can be thought of as $-0.75 + j0$, giving a phase angle of $180°$.)

3.17 UNSTABLE SYSTEMS

Don't forget that discrete systems with transfer function poles *outside* the unit circle are unstable, and so must be redesigned. Figure 3.30(a) shows the unit step response for a system with a pole at 1.2, while Fig. 3.30(b) is the unit step response for a filter with conjugate poles at $0.8 \pm j0.8$, again, outside the unit circle. More work on these particular designs is clearly needed!

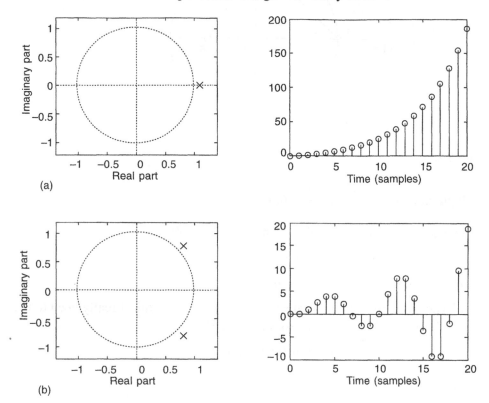

Figure 3.30

3.18   SELF-ASSESSMENT TEST

A digital filter has a transfer function given by

$$T(z) = \frac{z}{(z^2 + z + 0.5)(z - 0.8)}$$

The sampling frequency is 16 Hz.

(a) Plot the pole–zero diagram for the filter and, *hence,* find the gain and phase angle at 0 Hz and 4 Hz.

(a) Check your gain and phase values at 4 Hz *directly from the transfer function.*

3.19   RECAP

• The frequency response of a continuous system can be found by replacing *s* with $j\omega$ in the transfer function and, for a discrete system, by replacing z with $e^{j\omega T}$.

- The frequency response can also be found from the transfer function p–z diagrams using

$$\text{Gain} = k\,\frac{\Pi \text{ zero distances}}{\Pi \text{ pole distances}}$$

and

$$\text{Phase angle} = \Sigma \text{ zero angles} - \Sigma \text{ pole angles}$$

- The frequency point moves up the imaginary axis from the origin (d.c.) for continuous systems and around the unit circle in an anticlockwise direction from $z = 1$ (d.c.) to $z = -1$ (the Nyquist frequency) for discrete systems.

3.20 CHAPTER SUMMARY

We started this chapter by looking at the s-plane – i.e. by reviewing how continuous signals and systems can be analysed from their pole–zero diagrams. We then moved to the z-plane, which serves the equivalent purpose for discrete systems and signals. The link between the s- and z-planes was shown to be provided by $z = e^{sT}$.

The regions in the z-plane corresponding to stability, marginal stability and instability were identified as being *within* the unit circle, *on* the unit circle and *outside* the unit circle respectively.

The shapes of discrete signals were established from the positions of their poles and zeros – the general shape being determined by the pole positions, while the zero positions govern the magnitude of the signals. The presence or otherwise of signal zeros also controls signal delays.

The frequency response of a discrete system was derived directly from the z-plane transfer function by using $z = e^{j\omega T} = \cos \omega T + j \sin \omega T$. The frequency response was also obtained from the p–z diagram of the system transfer function using

$$\text{Gain} = k\,\frac{\Pi \text{ zero distances}}{\Pi \text{ pole distances}}$$

and

$$\text{Phase angle} = \Sigma \text{ zero angles} - \Sigma \text{ pole angles}$$

This chapter has been mainly concerned with the analysis of discrete systems but you have now also acquired many of the tools needed to *design* digital systems.

3.21   PROBLEMS

N.B.   If you have access to MATLAB then use it wherever possible to check your answers to the following problems.

1   A filter has a transfer function of $T(z) = (z - 0.1)/(z - 0.3)$ and is used with a sampling frequency of 1 kHz.

(a)   Sketch the pole–zero diagram for the filter.

(b)   Use your pole–zero diagram to estimate the gain and phase angle at 0 Hz and 250 Hz.

(c)   Check the gain and phase angle at 250 Hz *directly from the transfer function*. Also use this method to find the gain and phase angle at 375 Hz.

2   A continuous filter has a transfer function, $T(s)$, given by $T(s) = (s + 2)/[(s + 1)(s + 3)]$.

(a)   Design a digital filter that has equivalent poles and zeros by making use of the relationship $z = e^{sT}$. The sampling frequency is 10 Hz.

(b)   Compare the d.c. gains of the two filters either with the aid of their p–z diagram or directly from their transfer functions. Comment on these values.

3   A digital filter has the z-domain transfer function of $1/(z^2 - z + 0.5)$. Sketch the p–z diagrams for the filter outputs when the input is (a) a unit sample sequence and (b) a discrete unit step. Use the system p–z diagram to find the gain of the system at 0 Hz and also at $f_s/4$, where f_s is the sampling frequency.

4   By using the function $z^2/(z + 1)$ as an example, explain why it is impossible for a system transfer function to have more zeros than poles. (Hint: Consider the response of the system to a unit sample sequence.)

5   A digital filter has a transfer function of the form $T(z) = (z^2 + 0.81)/(z^2 + p^2)$. The filter is to have a minimum gain of –6 dB occurring at a frequency of 50 Hz. Given that the sampling frequency is 200 Hz, find the appropriate p-value. (Hint: First sketch a p–z diagram to show the zero positions.) Also find the gain at 0 and 100 Hz. Hence sketch the *magnitude* response of the filter.

4 The design of IIR filters

4.1 CHAPTER PREVIEW

We are now in a position to consider the various approaches to the design of digital filters. The two types of filter, FIR and IIR, have very different design methods and so will be considered separately. In this chapter we will focus on the techniques used to design IIR or recursive filters. We begin with a very brief résumé of filter essentials.

4.2 FILTER BASICS

Before we move on to the design of digital filters, it is probably worth having a *very* brief recap on filters.

An *ideal* filter will have a constant gain of at least unity in the passband and a constant gain of zero in the stopband. Also, the gain should increase from the zero of the stopband to the higher gain of the passband at a single frequency, i.e. it should have a 'brick wall' profile. The magnitude responses of ideal lowpass, highpass, bandpass and bandstop filters are as shown in Fig. 4.1(a), (b), (c) and (d).

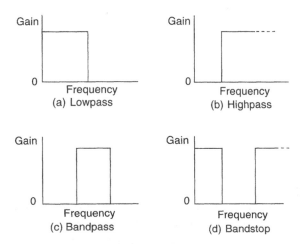

Figure 4.1

It is impossible to design a practical filter, either analogue or digital, that will have these profiles. Figure 4.2, for example, shows the magnitude response for a practical lowpass filter. The passband and stopband are not perfectly flat, the 'shoulder' between these two regions is very rounded and the transition between them, the 'roll-off' region, takes place over a wide frequency range. The closer we require our filter to agree with the ideal characteristics, the more complicated is the filter transfer function.

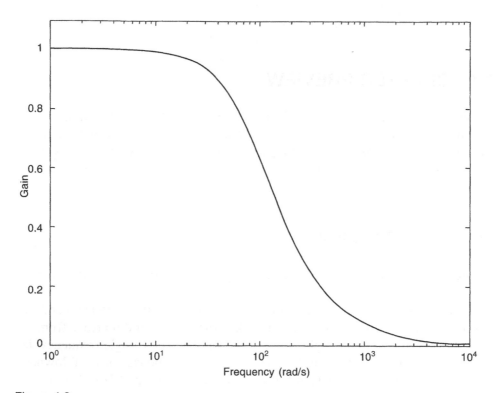

Figure 4.2

As the gain of a real filter does not drop vertically between the passband and stopband, we need some way of defining the 'cut-off' frequencies of filters, i.e. the effective end of the passband. The point chosen is the '−3 dB' point. This is the frequency at which the gain has fallen by 3 dB, or to $1/\sqrt{2}$ of its maximum value (gain in dB = $20 \log_{10}$ [gain]).

If you are rusty on the basic principles of analogue filters, this would be a good time to do some background reading. Some keywords to look for are: lowpass, highpass, bandstop, bandpass, cut-off frequency, roll-off, first, second (etc.) order, passive and active filters, Bode plots and dB. Howatson (1996) is just one of an abundance of circuit theory and analysis texts which will be relevant.

Much work has been carried out into the design of analogue filters and, as a result, standard design equations for analogue filters with very high specifications are available. However, as has been stressed earlier in this book, the characteristics of all analogue systems alter due to temperature changes and ageing. It is also impossible for two analogue systems to perform identically. Digital filters do not have these defects. They are also much more versatile than analogue filters in that they are programmable.

We will now look at various methods of designing digital filters.

4.3 FIR AND IIR FILTERS

You will remember from earlier chapters that digital filters can be divided broadly into two types – finite impulse response (FIR) and infinite impulse response (IIR) filters. If a single pulse is used as the input for an FIR filter the output pulses last for a finite time, while the output from an IIR filter will, theoretically, continue for ever. *Generally,* FIR filters are non-recursive, i.e. do not use feedback, while IIR do. It should be pointed out at this stage that it *is* possible to design some FIR filters which are recursive. However, for convenience, the term FIR filter will be taken to mean a non-recursive filter in this book.

The general expression for the transfer function, $T(z)$, of an FIR filter is:

$$T(z) = a_0 + a_1 z^{-1} + a_2 z^{-2} + a_3 z^{-3} \ldots$$

while that for the IIR filter is:

$$T(z) = \frac{a_0 + a_1 z^{-1} + a_2 z^{-2} + a_3 z^{-3} \ldots}{1 - b_1 z^{-1} - b_2 z^{-2} - b_3 z^{-3} \ldots}$$

Although they have the disadvantage of requiring more coefficients to achieve a similar filter performance, FIR filters do have the advantage that they will never be unstable, unlike IIR filters. FIR filters also have the potential to feature a linear phase response – but more on this later. The main point is that the two types of filter are very different in their performance *and also in their design.* In this chapter we will deal with the design of IIR filters only; FIR filters are considered in the next chapter.

There are several approaches that can be taken to the design of IIR filters. One common method is to take a standard analogue filter, e.g. Butterworth, Chebyshev, Bessell, etc., and convert this into its discrete equivalent. An alternative procedure is to start with the z-plane p–z diagram and attempt to place poles and zeros so as to produce the desired frequency response. This is often called the 'direct' design method. We will now look at these two approaches in detail.

4.4 THE DIRECT DESIGN OF IIR FILTERS

With this method, poles and zeros are placed on the z-plane in an attempt to achieve the required frequency response. When designing digital filters in this way there is sometimes an element of 'trial and error' involved. Often this is small and just serves to 'tweak' the calculated pole–zero positions, so as to improve the response, but at other times it is significant – this is usually when it proves very difficult to calculate the optimum pole–zero positions accurately. The reliance on trial and error has obviously been made much more viable with the increased availability of CAD packages such as MATLAB. These packages allow us to simulate and then modify our designs very quickly.

Once a suitable p–z diagram has been established, the system transfer function can be derived. One of the many pleasures of using digital filters is that it is very easy to convert the transfer function into the actual filter, i.e. to convert it into the corresponding DSP software. This is far from true with analogue filters. Here, once we arrive at a suitable p–z diagram, the corresponding transfer function then needs to be converted into suitable *hardware* – i.e. electrical components. This is a much more difficult task!

This 'direct' method of IIR filter design is best described by means of examples, and so two follow.

Example 4.1 _____

A lowpass digital filter is required that has a d.c. gain of 1 and a cut-off frequency which is 0.25 of the sampling frequency. The filter is to have a transfer function of the form $T(z) = k(z + a)/(z + b)$.

Solution

Our first task is to locate the single zero and pole. To complete the design we then need to calculate a suitable k-value.

As we require a *lowpass* filter then, after the initial passband, the gain must fall as the frequency increases, i.e. as we move around the z-plane unit circle in an anticlockwise direction, starting at $z = 1$ (d.c.).

Remembering that

$$\text{Gain} = k \frac{\Pi \text{ zero distances}}{\Pi \text{ pole distances}}$$

it would therefore be sensible to arrange for the zero distance to be zero at the Nyquist frequency. In other words, we will need to place the filter zero at $z = -1$.

$$\therefore a = +1$$

Common sense suggests that, in order to achieve the required response, the single pole probably needs to be placed somewhere on the *positive* real axis, i.e. this will ensure a large gain for low frequencies and a lower gain for high frequencies. Figure 4.3 shows the zero at $z = -1$ and the pole at $z = d$. Point A corresponds to 0 Hz and point B to the cut-off frequency, which is half the Nyquist frequency (a quarter of the sampling frequency).

Using

$$\text{Gain} = k \frac{\Pi \text{ zero distances}}{\Pi \text{ pole distances}}$$

the gains at 0 Hz and $0.5f_N$ are given by:

$$\text{Gain}_A = \frac{k \times 2}{(1 - d)} \text{ and } \text{Gain}_B = \frac{k\sqrt{2}}{\sqrt{1 + d^2}}$$

$$\therefore \frac{\text{Gain}_B}{\text{Gain}_A} = \frac{k\sqrt{2}(1 - d)}{k2\sqrt{1 + d^2}}$$

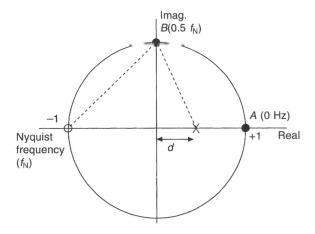

Figure 4.3

But *B* is the −3 dB point, therefore

$$\frac{\text{Gain}_B}{\text{Gain}_A} = \frac{1}{\sqrt{2}}$$

$$\therefore \frac{1}{\sqrt{2}} = \frac{k\sqrt{2}(1-d)}{k2\sqrt{1+d^2}}$$

$$\therefore \frac{1}{2} = \frac{2(1-d)^2}{4(1+d^2)}$$

$$\therefore 1 + d^2 = 1 - 2d + d^2$$

$$\therefore 1 = 1 - 2d$$

$$\therefore d = 0$$

Therefore the pole must be placed at the origin.

$$\therefore b = 0$$

Applying

$$\text{Gain} = k\frac{\Pi \text{ zero distances}}{\Pi \text{ pole distances}}$$

at point A:

$$1 = \frac{k \times 2}{1}$$

$$\therefore k = 0.5$$

Substituting the *a*-, *b*- and *k*-values into the transfer function gives us:

$$T(z) = \frac{0.5(z+1)}{z}$$

Figure 4.4 shows that the −3 dB point *does* occur at 0.5f_N and that the d.c. gain is 1 (0 dB), and so our design has satisfied the filter specification.

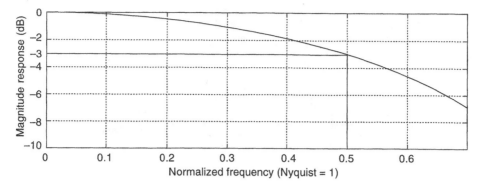

Figure 4.4

Example 4.2

A digital *notch filter* is required which has a notch frequency of 20 Hz, a bandwidth of no more than 4 Hz and an attenuation at the notch frequency of at least 40 dB. The gain in the passband is to be 1. The sampling frequency is 160 Hz.

Solution

A notch filter is a very narrow bandwidth bandstop filter – it has a gain which drops and then rises very steeply with increasing frequency and so the magnitude response has the shape of a notch.

As

$$\text{Gain} = k\frac{\Pi \text{ zero distances}}{\Pi \text{ pole distances}}$$

it follows that we must make the 'zero distance' very small at the notch frequency. As the specification stipulates that the attenuation needs to be *at least* 40 dB then the easiest thing to do is to place a zero *on* the unit circle at the position corresponding to the notch frequency of 20 Hz. This, theoretically, should cause the gain to fall to zero. This z-plane zero will, of course, be one of two complex conjugate zeros.

As the sampling frequency is 160 Hz then the Nyquist frequency is 80 Hz and so the notch frequency of 20 Hz corresponds to an angle of 45° on the z-plane diagram. Figure 4.5 shows the zero, Z, along with its complex conjugate zero, placed on the unit circle at the position corresponding to this frequency.

We now need to place the complex conjugate poles so that we obtain the correct bandwidth of no more than 4 Hz. In other words, the gain must fall to –3 dB, or to approximately $1/\sqrt{2}$ of its passband gain, at 18 Hz and 22 Hz.

From the symmetry of the frequency response, it seems sensible to place the poles on the same radii as the zeros. The two poles will have to be sufficiently close to the corresponding zeros to ensure that the gain is approximately unity for all of the frequency range, apart from the 'notch'. As they are so close then $k = 1$ is required to give a passband gain of 1. Their locations must also be such that the bandwidth is 4 Hz. We *might* find that this specification is impossible to achieve.

In Figure 4.5 the two poles are shown as being a distance d away from the two zeros. Also shown are the two –3 dB points, *B* and *A*, at 18 Hz and 22 Hz respectively (not to scale)

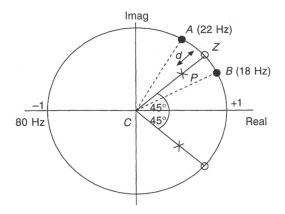

Figure 4.5

The angle *ACB* corresponds to 4 Hz and so, as $f_N = 80$ Hz,

$$\angle ACB = \frac{4}{80}\pi \approx 0.16 \text{ rad}$$

Using $s = r\theta$, where *s* is the arc of a circle, *r* the radius, and θ the angle subtended by the arc, then, as $r = 1$, the arc length from *A* to *B* must also be 0.16, and so $ZA = ZB = 0.08$.

As *A* is very close to *Z*, the pole, the zero and *A* (or *B*) form an approximate right-angled triangle, with $\angle AZP \approx 90°$. Therefore the pole distance from *A* or *B* must be very close to $\sqrt{d^2 + 0.08^2}$.

As

$$\text{Gain} = k\frac{\Pi\,\text{zero distances}}{\Pi\,\text{pole distances}}$$

and $k = 1$,

$$\frac{1}{\sqrt{2}} = \frac{AZ}{AP}$$

$$\therefore \frac{1}{\sqrt{2}} = \frac{0.08}{\sqrt{d^2 + 0.08^2}}$$

$$\therefore (d^2 + 0.08^2) = 2 \times 0.08^2$$

$$\therefore d = 0.08$$

Therefore, the poles must be 0.92 from the origin.

It follows that we must place the zeros at $\cos 45° \pm j \sin 45°$, i.e. at $0.707 \pm j0.707$, and the poles at $0.92 \cos 45° \pm j\,0.92 \sin 45°$ or $0.651 \pm j\,0.651$.

$$\therefore T(z) = \frac{(z - 0.707 + j0.707)(z - 0.707 - j0.707)}{(z - 0.651 + j0.651)(z - 0.651 - j0.651)} = \frac{z^2 - 1.414z + 1}{z^2 - 1.302z + 0.847}$$

N.B.   The effect of the other pole and zero has not been taken into account in the calculation. This is
because they are so close together that they tend to cancel each other out and so should not affect
the gain significantly.

The design is now complete; all that remains is to check the response – the
MATLAB plot is shown in Fig. 4.6. It can be seen that the magnitude response
satisfies the specification, with the notch frequency occurring at $0.25f_N$ (20 Hz),
the notch attenuation being at least 40 dB and the bandwidth approximately
$0.05f_N$ (4 Hz).

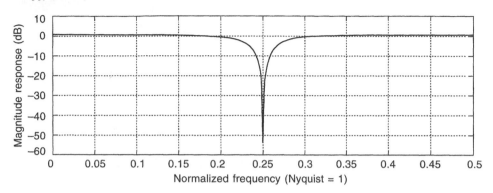

Figure 4.6

If we had found that the response was not satisfactory then the pole positions
could easily be 'tweaked', with the help of MATLAB, in order to improve the
performance.

Question
Would it be possible to design the filter using zeros *only*?

Answer
A resounding *no*!

This approach has a fatal flaw and is doomed to failure. Although,
theoretically, it might be possible to place zeros on the p–z diagram so as to
satisfy the magnitude response, *practically* it is impossible for the transfer
function of a discrete system to have more zeros than poles. Such a system
would produce an output signal before it receives an input signal! If you're
not convinced see problem 4 at the end of Chapter 3.

N.B.   We probably overdid the calculation in these examples and, practically, could have relied more on
MATLAB to decide on the location of the poles and zeros. For example, once we had calculated
the zero positions in the 'notch' problem, we could have found suitable positions for the poles by
trial and error, which would have saved a lot of work. However, it is important that you understand
the underlying principles in the design process, and so the extra calculations were worthwhile –
no doubt you feel much better for having done it?

4.5   SELF-ASSESSMENT TEST

A discrete bandpass filter is required that has a centre frequency of 100 Hz, a bandwidth of 20 Hz and a peak gain of 1. The filter is to have a transfer function of the form $T(z) = k/(z^2 + a^2)$ and the sampling frequency is 400 Hz.

4.6 RECAP

- The direct method for the design of IIR filters relies on poles and zeros being placed at suitable points in the z-plane so as to achieve an adequate frequency response.

- Some calculation is usually needed for the initial location of the poles and zeros but the final fine-tuning can often be done with the help of suitable software, i.e. by using 'trial and error'.

4.7 THE DESIGN OF IIR FILTERS VIA ANALOGUE FILTERS

The direct design method can give reasonable results for fairly simple filters. However, it is far from ideal when something rather more sophisticated is needed, and so an alternative approach is required.

A huge amount of research has been carried out in the past into the design of analogue filters and, as a result, 'classic' filters such as Butterworth, Chebyshev, 'elliptic' etc. have been developed, each one having its own particular good and less good features. A modern-day designer of analogue filters therefore has an excellent library of designs at his or her disposal. It would be a dreadful waste of effort not to try to make use of this expertise when designing *digital* filters. Very sensibly, various s-domain to z-domain transforms have been developed which allow us to convert analogue filters into their digital 'equivalents' fairly easily. It must always be remembered that all transforms are approximations and that *no* digital filter can be identical to its analogue prototype, i.e. have exactly the same magnitude, phase and time responses. The various s-to-z transformation methods have advantages and disadvantages. Remember also that a digital filter should not be expected to operate above the Nyquist frequency – this is clearly not a consideration when building analogue filters.

There are many standard methods of converting an analogue filter to its digital equivalent – a process called 'discretization'. We will look at three of the most common – the *bilinear* transform, the *impulse-invariant* method and the *pole–zero matching* technique.

4.8 THE BILINEAR TRANSFORM

I mentioned earlier that there are classic analogue filters which we can use as templates when we need to design a digital filter. However, some of these have fairly complicated s-domain transfer functions and so, to save having to deal with unnecessarily complicated algebra at this stage, a very simple filter will be used to demonstrate the various conversion methods. The prototype filter that will be used is a first-order, lowpass filter having the transfer function:

$$T(s) = \frac{1}{1 + s/\omega_c}$$

(The filter is classed as a *first-order* filter as the highest power of s is 1. Generally, the higher the order the better the filter performance.)

This transfer function defines a lowpass filter with a cut-off angular frequency of ω_c rad/s. For example, if we choose $\omega_c = 10$ rad/s, then:

$$T(s) = \frac{1}{1 + s/10}$$

or

$$T(s) = \frac{10}{s + 10}$$

Figure 4.7 confirms the predicted frequency response of the filter, i.e. it *is* a lowpass filter with a cut-off frequency of 10 rad/s (or approximately 1.6 Hz).

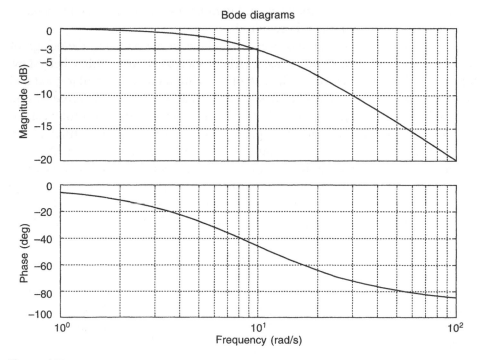

Figure 4.7

To apply the *bilinear* transform we have to replace s with:

$$s = \frac{2(z - 1)}{T(z + 1)}$$

where T is the sampling period. (See Virk (1991), Oppenheim and Schafer (1975) or Ludeman (1987) for an explanation of the origin of this transform.)

We will now convert the analogue lowpass filter to its discrete equivalent using a sampling frequency of 16 Hz ($T = 0.065$ s).

$$\therefore T(z) = \frac{10}{\dfrac{2}{0.0625}\dfrac{(z-1)}{(z+1)} + 10}$$

$$= \frac{10(z+1)}{32(z-1) + 10(z+1)}$$

$$= \frac{10(z+1)}{42z - 22}$$

$$= \frac{10(z+1)}{42(z - 0.524)}$$

$$\therefore T(z) = \frac{0.238(z+1)}{(z - 0.524)}$$

The frequency response for this filter is shown in Fig. 4.8.

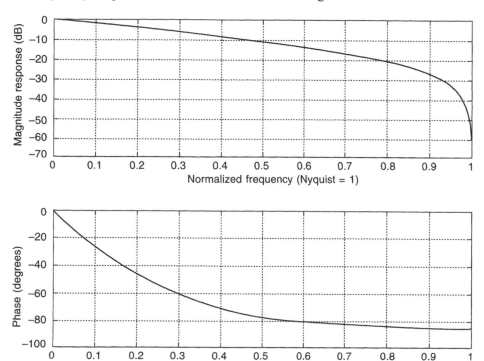

Figure 4.8

Taking into account the linear frequency scale, the phase response has roughly the same shape as that of the analogue filter, while the magnitude response appears to be in very good agreement – it is certainly a lowpass filter! However, a more detailed examination reveals that the cut-off frequency is approximately $0.19f_N$, or 1.5 Hz, rather than the 1.6 Hz ($0.2f_N$) of the analogue filter, which is an error of approximately 6%. This is a recognized problem associated with the bilinear transform, i.e. breakpoint frequencies might be changed significantly during the transformation.

However, all is not lost. If the approximation is too poor to be acceptable, then we need to '*pre-warp*' the frequencies before we apply the bilinear transformation. This entails modifying the s-domain transfer function in a particular way, such that the gain at a 'frequency of interest' (usually an important breakpoint) is the same for the discrete filter as the continuous filter. (N.B. Frequency pre-warping does not change the d.c. gain.)

If the frequency at which we want to equate the gains is ω_c, then we first need to calculate a 'pre-warped' version, ω'_c, where:

$$\omega'_c = \frac{2}{T} \tan\left(\frac{\omega T}{2}\right)$$

In this example we have just one frequency which is of obvious interest – the single breakpoint of 10 rad/s.

$$\therefore \omega'_c = 32 \tan\left(\frac{10}{32}\right) \approx 10.34$$

We now need to redesign the digital filter starting with the 'pre-warped' s-domain transfer function. There are several ways of doing this, but the easiest is to first change the format of the transfer function such that every s is replaced with s/ω_c.

So, dividing numerator and denominator by 10, our simple transfer function now becomes:

$$T(s) = \frac{1}{\dfrac{s}{10} + 1}$$

The final step is to replace all s/ω_c terms with s/ω'_c, i.e. with $s/10.34$ in our example.

$$\therefore T(s)' = \frac{1}{\dfrac{s}{10.34} + 1}$$

$$\therefore T(s)' = \frac{10.34}{s + 10.34}$$

We now apply the bilinear transformation to this pre-warped version of the transfer function. This gives:

$$T(z) = \frac{0.244(z + 1)}{z - 0.512} \quad \text{(check this for yourself)}$$

Figure 4.9 shows the magnitude response before and after pre-warping. The normalized frequency scale has been expanded so as to make the difference between the responses more obvious. The new cut-off frequency now corresponds much more closely to the $0.2f_N$ or 1.6 Hz required.

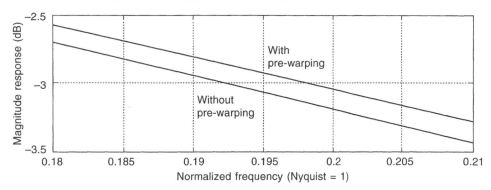

Figure 4.9

N.B.1   It is always worth checking whether the frequency needs to be pre-warped before carrying out the bilinear transformation. If ω_c' differs from ω_c by more than about 1%, then it is probably worth pre-warping.

N.B.2   If we had used a sufficiently high sampling frequency in the example above, then pre-warping would not have been necessary. For example, if we had chosen f_s to have been 100 Hz rather than 16 Hz, then $\omega_c' = 10.008$ rad/s, which is not significantly different from ω_c, and so it is probably not worth pre-warping.

N.B.3   Pre-warping will only ensure that the gain is matched at the chosen frequency (and 0 Hz). If there had been several breakpoints, for example, if

$$T(z) = \frac{s + 2}{(s + 3)(s + 4)}$$

then a decision has to be made as to which of the three breakpoints it is best to preserve. An example follows.

Example 4.3

A continuous filter has a transfer function of

$$T(s) = \frac{2}{(s + 1)(s + 2)}$$

It is to be converted to its digital equivalent using the bilinear transform. It is important that the gain is preserved at the upper breakpoint frequency of 2 rad/s. The sampling frequency used is 2 Hz.

Solution

We first need to check whether pre-warping is necessary to maintain the breakpoints. Using

$$\omega_c' = \frac{2}{T} \tan\left(\frac{\omega T}{2}\right)$$

$$\omega_c' = \frac{2}{0.5} \tan\left(\frac{2 \times 0.5}{2}\right) = 2.19$$

As the difference between ω_c' and ω_c is significant (about 10%), then pre-warping is necessary.

Changing the format of the transfer function by replacing all s-values with s/ω_c, i.e. $s/2$, gives:

$$T(s) = \frac{0.5}{\left(\frac{s}{2} + 0.5\right)\left(\frac{s}{2} + 1\right)}$$

(Here we have divided both bracketed terms by 2, in other words we have divided the denominator by 4. We therefore also need to divide the numerator by 4 – hence the '0.5'.)

Replacing all $s/2$ terms with $s/2.19$:

$$T(s)' = \frac{0.5}{\left(\frac{s}{2.19} + 0.5\right)\left(\frac{s}{2.19} + 1\right)}$$

$$\therefore T(s)' \approx \frac{2.4}{(s + 1.1)(s + 2.19)}$$

Applying the bilinear transform, i.e. replacing s with $4(z - 1)/(z + 1)$, results in the z-domain transfer function of:

$$T(z) = 0.073 \frac{(z + 1)^2}{(z - 0.58)(z - 0.307)}$$

4.9   SELF-ASSESSMENT TEST

Convert the filter with the transfer function of $T(s) = 16/(s + 4)^2$ into its discrete equivalent, using the bilinear transformation. The discrete filter is to be used with a sampling frequency of 5 Hz.

4.10   THE IMPULSE-INVARIANT METHOD

The approach here is to produce a discrete filter that has a *unit sample response* which is 'the same' as the *unit impulse response* of the continuous prototype, as shown in Fig. 4.10. Clearly, the output from the digital data processor will be a sampled response, however, if properly designed, the *envelope* of the signal should be the same as the unit impulse response of the analogue filter.

Using our prototype lowpass filter as an example, we start with:

$$T(s) = \frac{10}{s + 10}$$

If we use a unit impulse as the input then the Laplace transform of the output, i.e. the 'unit impulse response', is also $10/(s + 10)$. This is because the Laplace transform of a unit impulse is '1' and so $Y(s) = T(s) \times 1$.

If we now find the *inverse* Laplace transform, $y(t)$, then this tells us the shape of the unit impulse response in the time domain.

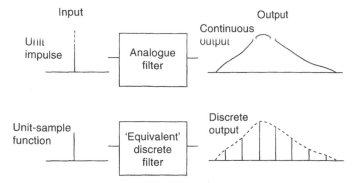

Figure 4.10

From the tables, the inverse Laplace transform of $1/(s + a)$ is e^{-at}, which is an exponentially decaying d.c. signal.

$$\therefore y(t) = 10e^{-10t}.$$

The final step is to find the z-transform, $Y(z)$, of this time variation. Again, from the Laplace/z-transform tables, e^{-at} has *a* z-transform of $z/(z - e^{-aT})$. As the sampling frequency is 16 Hz,

$$Y(z) = \frac{10z}{z - e^{-0.625}} = \frac{10z}{z - 0.535}$$

This function must also be the z-domain transfer function, $T(z)$, of our equivalent discrete filter. This is because the z-transform of a unit sample sequence is also 1, and so $Y(z) = T(z) \times 1$. In other words, if we inputted a unit sample sequence into a discrete filter having a transfer function of $10z/(z - 0.535)$, the z-transform of the output would also be $10z/(z - 0.535)$. This is exactly what we require of the discrete filter.

$$\therefore T(z) = \frac{10z}{z - 0.535}$$

To make sure that no mistakes have been made during the transformation, it's worth checking the unit impulse response and unit sample response of the continuous and discrete filters, respectively. Figure 4.11 shows the impressively close agreement between them – the continuous response plot passing through the circles of the discrete response 'stems', so confirming that the transformation has been carried out correctly.

Figure 4.12 shows its frequency response – this doesn't look quite so good! Comparing with Fig. 4.7, the phase response is clearly very different. The phase response *is* important but not as much as the magnitude response – we will discuss phase response more in the next chapter and will concern ourselves solely with the magnitude response here. Although the –3 dB point on the magnitude response is close to 0.2 f_N, i.e. the normalized frequency which corresponds to 1.6 Hz, the gain values are *much* bigger than we might have expected. However, we could have predicted this gain 'mismatch' from an examination of the discrete filter's p–z diagram, Fig. 4.13.

agreement with the required response, especially well away from the Nyquist frequency of approximately 50 rad/s.

N.B. 1   Ideally we should sample at a frequency which is much higher than the desired cut-off frequency of 1.6 Hz. We are sampling at 10× this frequency which is adequate. However, if we had reduced the sampling frequency down to 10 Hz, for example, then this would have resulted in the –3 dB point occurring at 1.75 Hz, which is a significant error.

N.B. 2   During the design process we used the Laplace/z-transform tables to move from the s-domain to the z-domain, via the time domain. Clearly, there is no need to actually enter the time domain as the Laplace/z-transform tables allow us to convert *directly* from the s- to the z-domain, i.e. $1/(s + a) \equiv z/(z - e^{-aT})$.

Example 4.4

An analogue filter has the transfer function, given by:

$$T(s) = \frac{6(s + 1)}{(s + 2)(s + 3)}$$

Convert this to its discrete equivalent by using the impulse-invariant transform. The sampling frequency is 20 Hz.

Solution

The first thing to do is to try to identify a function of this 'shape' in the Laplace transform tables. Unfortunately this one doesn't appear. What we must therefore do is break it down into functions that *do* appear, by using the method of partial fractions, i.e.:

$$T(s) = \frac{6(s + 1)}{(s + 2)(s + 3)} = \frac{A}{s + 2} + \frac{B}{s + 3}$$

By using the 'cover-up rule', or any other method, you should find that $A = -6$, and $B = 12$,

$$\therefore T(s) = \frac{12}{s + 3} - \frac{6}{s + 2}$$

Using the Laplace/z-transform tables to convert to the z-plane gives us

$$T(z) = \frac{12z}{z - e^{-3 \times 0.05}} - \frac{6z}{z - e^{-2 \times 0.05}} = \frac{12z}{z - 0.86} - \frac{6z}{z - 0.905}$$

$$\therefore T(z) = \frac{12z(z - 0.905) - 6z(z - 0.86)}{(z - 0.86)(z - 0.905)} = \frac{z(12z - 10.86 - 6z + 5.16)}{(z - 0.86)(z - 0.905)}$$

$$\therefore T(z) = \frac{6z(z - 0.95)}{(z - 0.86)(z - 0.905)}$$

We now need to make the gains of the two the same at 0 Hz. We could do this by first plotting the two p–z diagrams for the filters and then finding the gains of the two filters at 0 Hz from the pole and zero distances. However, there is no need to do this as we can derive the d.c. gains of the two filters *directly* from their transfer functions.

Remembering that $s = 0$ and $z = 1$ correspond to a frequency of 0 Hz in the s-domain and z-domain respectively, we require to find the value of k, such that:

$$\left| \frac{6(s + 10)}{(s + 2)(s + 3)} \right|_{s=0} = k \left| \frac{6z(z - 0.95)}{(z - 0.86)(z - 0.905)} \right|_{z=1}$$

$$\therefore \frac{60}{6} = k \frac{6 \times 0.05}{0.14 \times 0.095}$$

$$\therefore k = 0.443$$

$$\therefore T(z) = \frac{2.66z(z - 0.95)}{(z - 0.86)(z - 0.905)}$$

4.11 SELF-ASSESSMENT TEST

Two analogue filters have the following transfer functions. Convert them to their digital equivalents, using the *impulse-invariant transformation.*

(a) $T(s) = \dfrac{4}{(s + 2)^2}$

(b) $T(s) = \dfrac{2}{(s + 1)(s + 2)}$

Assume a sampling frequency of 10 Hz.

4.12 POLE–ZERO MAPPING

The approach adopted here is to use the relationship $z = e^{sT}$ to convert the s-plane poles and zeros of the continuous filter to equivalent poles and zeros in the z-plane. Once the z-plane poles and zeros are known, the general form of the z-domain transfer function can be derived. All that then remains to be done is to ensure that the two filters have the same gain, at least at some important frequency – often 0 Hz.

Once again we will use our prototype lowpass filter, with the transfer function of $T(s) = 10/(s + 10)$, to illustrate this method, along with a sampling frequency of 16 Hz.

This particular filter has just a single pole at $s = -10$. So, using $z = e^{sT}$:

$$z = e^{-10 \times 0.0625} = 0.535$$

It follows that our equivalent discrete filter has a z-domain transfer function with the general form of $T(z) = k/(z - 0.535)$.

We now need to find the value of k which will make the gains of the two filters the same at a particular frequency. As it is a lowpass filter, it makes sense to use 0 Hz.

$$\therefore \left| \frac{10}{s + 10} \right|_{s=0} = \left| \frac{k}{z - 0.535} \right|_{z=1}$$

$$\therefore 1 = \frac{k}{0.465}$$

$$\therefore k = 0.465$$

Therefore, the transfer function of our equivalent discrete filter is given by:

$$T(z) = \frac{0.465}{z - 0.535}$$

N.B.   Although this method is a very simple one, it generally gives poorer results than the two others, i.e. the bilinear transform and the impulse-invariant method. One disadvantage is that if the prototype continuous filter has no zeros then the discrete 'equivalent' will have no zeros. This is not true with the two other conversion methods. This might not seem to be a major problem but zeros *are* useful in that they help to shape the frequency response of the discrete filter. For example, they can increase the rate at which roll-off occurs. To get around this problem, zeros are often added to 'all-pole' filters so as to balance the number of poles – these zeros are usually placed at $z = -1$. This makes sense because all-pole analogue filters will be lowpass or bandpass filters (draw some pole diagrams to convince yourself of this) and so the equivalent digital filter must have a low gain at the Nyquist frequency.

Our filter design should therefore be improved if we add a single zero at $z = -1$. The transfer function now becomes:

$$T(z) = \frac{k(z + 1)}{z - 0.535}$$

For the d.c. gains of the continuous and discrete filters to be the same, we require that $k = 0.233$.

$$\therefore T(z) = \frac{0.233(z + 1)}{z - 0.535}$$

Occasionally the gains need to be equalized at a frequency other than 0 Hz. This is a bit more difficult and so an example is in order.

Let's say that we have an analogue filter with a transfer function of $s/(s + 1)$. If we convert this to its digital equivalent using the pole–zero mapping method and assuming a sampling frequency of 5 Hz, say, then you should find that the z-domain transfer function is given, approximately, by $T(z) = k(z - 1)/(z - 0.82)$.

> **Question**
> What type of filter is this?
>
> **Answer**
> It is a highpass filter. Looking at the s-domain transfer function, it's clear that the gain is zero when $s = 0$ and becomes 1 at sufficiently high frequencies, i.e. when $s \gg 1$. Similarly, the z-plane zero occurs at $z = 1$ (d.c.) and so the d.c. gain of the discrete filter must also be zero, and then increases as the frequency increases.

Anyway – back to the problem. It doesn't make very much sense to try to equalize gains at 0 Hz, as they are already the same, i.e. zero. Here we need to equalize the gains at a frequency in the passband. We could do this using MATLAB, or from the p–z diagram, or directly from the transfer function – the latter approach will be taken here.

Theoretically, the gain of the analogue filter never quite reaches 1. However, if we choose an angular frequency which is much greater than the cut-off frequency, then this should be well into the passband region. As the cut-off frequency is 1 rad/s then a frequency of 10 rad/s should be suitable. Substituting this value into the s-domain transfer function, i.e. replacing s with $j10$, gives a gain of 0.995, which is close enough to 1. We also need to check that this frequency is not above the Nyquist frequency. We are sampling at 5 Hz and so the Nyquist frequency is 2.5 Hz, or approximately 16 rad/s. It follows that our choice of 10 rad/s should be fine.

We therefore need to equalize the gains of the two filters when $s = j10$. Using $z = e^{sT}$, the corresponding z value is e^{j2}. Using Euler's identity, $z = \cos 2 + j \sin 2 = -0.416 + j0.91$.

$$\therefore \left| \frac{s}{s+1} \right|_{s=j10} = k \left| \frac{z-1}{z-0.82} \right|_{z=(-0.416+j0.91)}$$

$$\therefore \left| \frac{j10}{j10+1} \right| = k \left| \frac{-0.416 - 1 + j0.91}{-0.416 - 0.82 + j0.91} \right|$$

$$\therefore 0.995 \approx k \frac{1.683}{1.535}$$

$$\therefore k = 0.91$$

$$\therefore T(z) = 0.91 \frac{z-1}{z-0.82}$$

N.B. As this is a highpass filter, it would have made more sense to have just equalized the gains at the Nyquist frequency, i.e. at $s = j16$ and $z = -1$. This would have also been easier to do. However, if it were a bandpass filter, then we would have no choice but to equalize the gains at a frequency in the passband.

4.13 SELF-ASSESSMENT TEST

Convert the filters with the following transfer functions to their discrete equivalents, using the pole–zero mapping method. The gains of the digital filters are to be made the same as their analogue prototypes at 0 Hz. The digital filters are to be used with a sampling frequency of 10 Hz.

(a) $T(s) = \dfrac{4}{(s+2)^2}$ (add two z-plane zeros at $z = -1$)

(b) $T(s) = \dfrac{s+2}{(s+1)(s+3)}$

4.14   MATLAB AND *s*-TO-*z* TRANSFORMATIONS

Very conveniently, MATLAB has special functions that will perform the main *s*-to-*z* transformations for us. This 'c2dm' function will automatically generate the appropriate *z*-domain transfer function coefficients.

For example, if we wish to check our *s*-to-*z* conversion of the simple lowpass filter ($T(s) = 10/(s + 10)$) using the pole–zero mapping method, say, then the MATLAB program is:

```
n=[10];
d=[1 10];
T=0.0625;                 % sampling period
c2dm(n,d,T,'matched')     % the 'matched' indicates the
                          % p-z matching method
```

The bilinear transform, both with and without pre-warping, can also be performed. The bilinear transform can also be achieved using the 'bilinear' function, while the 'impinvar' function can be used to accomplish the impulse-invariant transformation. Consult the MATLAB user manual for further details.

4.15   CLASSIC ANALOGUE FILTERS

So far we have looked at how to convert some fairly simple analogue filters to their discrete equivalents. As mentioned earlier, there are many standard analogue filter types available to us as prototype filters. In other words, if we need to design a digital filter with a certain maximum ripple in the stopband and passband, a particular rate of roll-off etc. we do not have to start from scratch. All we need to do is select one of the classic filters, e.g. Butterworth, Chebyshev, etc., which best satisfies the specification and then use a suitable *s*-to-*z* transformation method to convert it to the equivalent digital filter.

It is not the aim of this text to cover the very important, but lengthy topic of analogue filters – here we are only concerned with how to convert a suitable analogue filter to its digital equivalent. If you have not met the standard filter types then you are advised to do some background reading. There are many books available on this topic but you will probably find Ingle and Proakis (1997), Steiglitz (1995) and also Denbigh (1998) particularly helpful as these texts also deal with the conversion of standard analogue filters to their digital equivalent. (Keywords: Butterworth, Chebyshev, elliptic, Bessell, pass/stopband ripple.)

I would not want to leave this particular approach to digital filter design without converting one of the classic prototypes to its discrete equivalent. The following worked example is based on a Butterworth filter. The main advantage of this particular family of filters is that they have very flat stop- and passbands. The major disadvantage is that the initial roll-off rate is fairly slow. In other words, the 'shoulder' between the pass and stop regions is rather rounded compared to some other filter types such as the members of the Chebyshev family. Chebyshev filters, in turn, have the drawback of a relatively large ripple in the passband or the stopband.

Example 4.5_____

A second-order, lowpass Butterworth filter has the following transfer function:

$$T(s) = \frac{1}{1 + \sqrt{2}\left(\dfrac{s}{\omega_c}\right) + \left(\dfrac{s}{\omega_c}\right)^2}$$

where ω_c is the cut-off angular frequency.

Using this as a prototype, design a lowpass *digital* filter, having a cut-off frequency of 70 rad/s. The bilinear transform is to be used for the conversion and the filter is to be used with a sampling frequency of 100 Hz.

Solution

We first need to check whether pre-warping is necessary.
Using $\omega_c' = (2/T)\tan(\omega T/2)$:

$$\omega_c' = \frac{2}{10^{-2}}\tan\left(\frac{70 \times 10^{-2}}{2}\right) = 73 \text{ rad/s}$$

This is significantly different from ω_c and so pre-warping *is* worthwhile.
Replacing ω_c with ω_c' in the filter transfer function:

$$T(s)' = \frac{1}{1 + \sqrt{2}\left(\dfrac{s}{73}\right) + \left(\dfrac{s}{73}\right)^2}$$

Using the bilinear transformation, i.e. $s = 2(z-1)/T(z+1)$:

$$s = 200\left(\frac{z-1}{z+1}\right)$$

$$\therefore T(z) = \frac{1}{1 + \dfrac{\sqrt{2} \times 200}{73}\left(\dfrac{z-1}{z+1}\right) + \left(\dfrac{200}{73}\right)^2\left(\dfrac{z-1}{z+1}\right)^2}$$

$$= \frac{1}{1 + 3.87\left(\dfrac{z-1}{z+1}\right) + 7.5\left(\dfrac{z-1}{z+1}\right)^2}$$

$$= \frac{(z+1)^2}{(z+1)^2 + 3.87(z-1)(z+1) + 7.5(z-1)^2}$$

$$= \frac{(z+1)^2}{z^2 + 2z + 1 + 3.87z^2 - 3.87 + 7.5z^2 + 7.5 - 15z}$$

$$\therefore T(z) = \frac{(z+1)^2}{12.37z^2 - 13z + 4.63}$$

Figure 4.15 compares the digital and analogue gain responses. As expected, the filter responses agree closely at lower frequencies but begin to diverge as the Nyquist frequency of 50 Hz (314 rad/s) is approached.

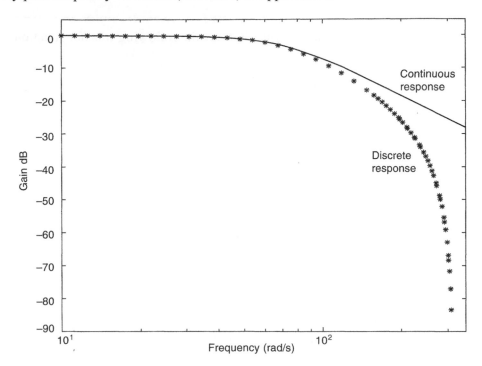

Figure 4.15

4.16 FREQUENCY TRANSFORMATION IN THE s-DOMAIN

You have probably noticed that all but one of the discrete filters we have designed, so far, have been lowpass filters. The reason for concentrating on this particular class of filter is that the transfer function for a lowpass filter can be used as a template for the design of other types, i.e. highpass, bandpass and bandstop. In other words, if we know the expression for the transfer function of a second-order, lowpass, Chebyshev filter, for example, then we can use it to derive the transfer function for a second-order, bandpass, Chebyshev filter. This is because there are standard transformations (yet again!) which can be used to convert the transfer function of a lowpass filter into the transfer function for any other filter type within that family of filters.

For example, if we wish to convert a lowpass filter with a cut-off angular frequency of ω_c to a *highpass* filter with a cut-off angular frequency of Ω_c, then we just need to replace s, in the s-domain transfer function for the lowpass filter with:

$$s = \frac{\omega_c \Omega_c}{s}$$

To demonstrate this we will use our old friend, the simple lowpass filter with the transfer function of $T(s) = 10/(s + 10)$. This lowpass filter has a cut-off frequency of 10 rad/s. Let's imagine that we want to design a simple first-order, highpass filter with a cut-off frequency of 6 rad/s.

We therefore need to replace s with $10 \times 6/s$, to produce our transfer function, $T_H(s)$, for the highpass filter.

$$\therefore T_H(s) = \frac{10}{\dfrac{60}{s} + 10}$$

$$\therefore T_H(s) = \frac{10s}{60 + 10s} = \frac{s}{s + 6}$$

The magnitude response, Fig. 4.16, is clearly that of a highpass filter with a cut-off angular frequency of 6 rad/s.

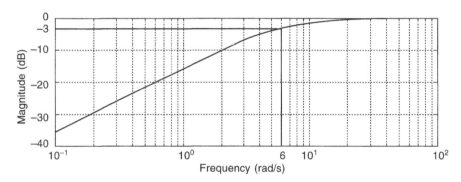

Figure 4.16

We can save ourselves much hard work by using the MATLAB functions lp2lp, lp2hp, lp2bp and lp2bs to convert the transfer function of a continuous lowpass filter into that for another lowpass or highpass, bandpass and bandstop filter, respectively.

For example, the MATLAB program

```
n=[10];              % numerator coefficient for
                     % the lowpass filter
d=[1 10];            % denominator - - - - -
wc=6;                % new cut-off frequency
[nh,dh]=lp2hp(n,d,wc) % convert to highpass transfer
                     % function
```

will generate the transfer function coefficients for our simple, first-order, highpass filter.

4.17 FREQUENCY TRANSFORMATION IN THE *z*-DOMAIN

In the above example, the conversion from lowpass to highpass filter was carried out in the *s*-domain. To produce the digital version of this we would then need to convert this highpass analogue filter to its digital equivalent using one of the various methods available to us, possibly with the help of MATLAB. However, if we know the *z*-domain transfer function for the discrete lowpass filter then this change from lowpass to highpass (or to any other type) *can also be done directly in the z-domain*.

To make this conversion from a lowpass to a highpass digital filter we have to replace *z* with:

$$z = -\frac{1 + az}{z + a}$$

where

$$a = -\frac{\cos\left[(\omega_c - \Omega_c)T/2\right]}{\cos\left[(\omega_c + \Omega_c)T/2\right]}$$

and ω_c and Ω_c are the cut-off angular frequencies for the low and highpass filters respectively.

As an example, we will transform the discrete equivalent of our lowpass filter with the *s*-domain transfer function $T(s) = 10/(s + 10)$. We will use the *z*-domain transfer function obtained earlier using the impulse-invariant technique (Section 4.10), i.e $T(z) = 0.465z/(z - 0.535)$. Again, a highpass filter with a cut-off frequency of 6 rad/s will be aimed for. The sampling frequency is 16 Hz ($T = 0.0625$ s).

From

$$a = -\frac{\cos\left[(\omega_c - \Omega_c)T/2\right]}{\cos\left[(\omega_c + \Omega_c)T/2\right]}$$

$$= -\frac{\cos\left[(10 - 6) \times 0.0625/2\right]}{\cos\left[(10 + 6) \times 0.0625/2\right]} = -1.13$$

Replacing *z* with $- (1 - 1.13z)/(z - 1.13)$ in $T(z) = 0.465z/(z - 0.535)$:

$$T_H(z) = \frac{0.465\left(\dfrac{1.13z - 1}{z - 1.13}\right)}{\left(\dfrac{1.13z - 1}{z - 1.13}\right) - 0.535}$$

$$= \frac{0.465(1.13z - 1)}{(1.13z - 1) - 0.535(z - 1.13)}$$

$$= \frac{0.465 \times 1.13(z - 0.885)}{0.595z - 0.395}$$

$$= \frac{0.465 \times 1.13(z - 0.885)}{0.595(z - 0.664)}$$

$$\therefore T_H(z) = \frac{0.88(z - 0.885)}{(z - 0.664)}$$

Figure 4.17 shows the frequency response of this filter. It *is* a highpass filter with a cut-off frequency which is very close to $0.12 f_N$, i.e. 0.96 Hz or 6 rad/s.

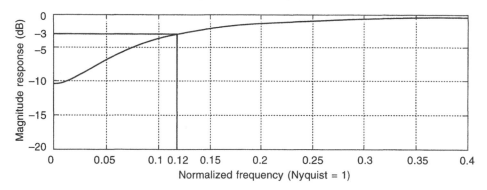

Figure 4.17

N.B.1   Although we have looked at a fairly simple example, *any* type of lowpass filter can be changed to another lowpass filter of different cut-off frequency or to equivalent highpass, bandpass or bandstop filters. The formulae for all conversions, both in the *s*-domain and the *z*-domain, are shown in Appendix B.

N.B.2   We have worked manually through the various steps of this particular approach to filter design, just relying on MATLAB to perform some stages of the design and also to check the performance of our filters. However, we can also use MATLAB to carry out the *complete* design for us. Once a decision has been made as to which family of filters is required, e.g. Butterworth, Chebyshev etc., the filter specification can then be entered and MATLAB will respond by generating the discrete filter transfer function coefficients *directly*.

For example, '[nz,dz] = butter(3, 0.4, 'high')' will generate the transfer function coefficients for a Butterworth, third-order, highpass filter, with a cut-off frequency equal to $0.4 \times f_N$, where f_N is the Nyquist frequency. If we add 'freqz(nz,dz)', to the program, the frequency response (Fig. 4.18) will also be plotted. This facility is obviously invaluable when it comes to the practical design of digital filters. Refer to the MATLAB user manual for further details.

4.18   SELF-ASSESSMENT TEST

A lowpass digital filter has a transfer function of $0.15\,(z+1)/(z-0.7)$. Given that this filter has a cut-off frequency of 55 Hz when sampled at a frequency of 1 kHz, use the table of Appendix B to convert it to a highpass filter with a cut-off frequency of 100 Hz. Assume a sampling frequency of 1 kHz.

4.19   RECAP

- A common way of designing IIR filters is to base the design on a suitable analogue filter. The prototype filter is often one of the tried and trusted classic filters such as Butterworth, Chebyshev etc. Each one of these filters has its own particular good and bad points and the initial choice of filter is a very important stage of the design process.

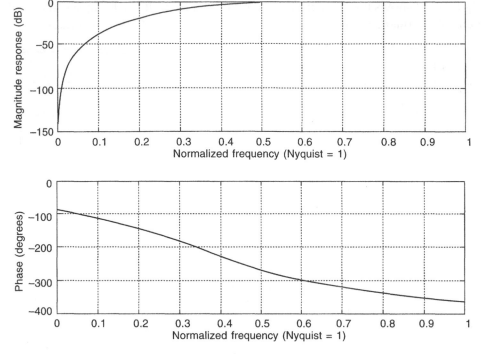

Figure 4.18

- Transformations such as the bilinear (often with pre-warping), pole–zero mapping and the impulse-invariant can then be used to convert the *s*-domain transfer function into its *z*-domain equivalent.

- Designs are usually based on a prototype of a lowpass filter. This is because the transfer function for all other types, i.e. highpass, bandpass, bandstop and lowpass filters with different cut-off frequencies can be derived from the transfer function of a lowpass filter by using yet more transformations.

- CAD packages are available which, once given the desired filter specification, will generate the transfer function of the digital filter for us directly.

4.20 PRACTICAL REALIZATION OF IIR FILTERS

Once the design process has been completed and simulation carried out to check the design, the final step is to program the DSP system, i.e. to write the controlling software. The essential data that must be entered into the program are the transfer function coefficients and the sampling frequency. In Chapter 2, the transfer function was derived for a general recursive filter. This has the form of:

$$T(z) = \frac{a_0 + a_1 z^{-1} + a_2 z^{-2} + a_3 z^{-3}}{1 - b_1 z^{-1} - b_2 z^{-2} - b_3 z^{-3} \ldots} \tag{4.1}$$

The transfer function coefficients must usually be entered into the DSP software in this form. For example, if we have a filter with a transfer function of:

$$T(z) = \frac{2z - 0.6}{5z^2 + 1.5z - 0.5}$$

then the format needs to be changed to:

$$T(z) = \frac{2z - 0.6}{5(z^2 + 0.3z - 0.1)}$$

$$= \frac{0.4z - 0.12}{z^2 + 0.3z - 0.1}$$

Dividing numerator and denominator by z^2 in order to change all indices to negative values and the initial value of the denominator to 1:

$$T(z) = \frac{0.4z^{-1} - 0.12z^{-2}}{1 + 0.3z^{-1} - 0.1z^{-2}}$$

Therefore, comparing with the general transfer function (equation (4.1)), $a_0 = 0$, $a_1 = 0.4$, $a_2 = -0.12$ and $b_1 = -0.3$, $b_2 = +0.1$ (*not* +0.3 and –0.1).

I should add a word of warning at this point. There *can* be unexpected problems with the filter performance if the coefficients are entered directly in this format. This is particularly likely when the sampling rate is very large compared to the filter breakpoints. This is because the filter poles and also the zeros tend to cluster together as f_s increases. In other words, the roots of the numerator and also the denominator polynomials become very close. To convince yourself of this, convert a filter with a transfer function of $(s + 2)/(s + 1)(s + 3)$ into its z-plane equivalent using the bilinear transform and a sampling frequency of 20 Hz, 200 Hz and 2000 Hz. You should find that the poles (and the zero) move closer and closer to $z = +1$ as the sampling frequency increases. Remember that, practically, there are a limited number of bits available in a processor to store each coefficient and so 'rounding' of coefficient values inevitably take place during processing. These approximation errors obviously become extremely significant when the roots of the polynomials are so close. A pole which should be near to the unit circle could even be calculated as being just outside – i.e. *disaster*! To overcome this 'clustering' problem, the filter is often broken down into smaller sections which are then cascaded to produce the complete filter. It can be shown that this structure reduces the problem. For more information on this method of implementation see Steiglitz (1995) and also Ingle and Proakis (1991, 1997). (Keywords: biquadratic sections.)

Hopefully, you are now in a position to do some practical designing and testing of IIR filters. Unfortunately, in a book of this length, it is not feasible to deal with the practical details of programming specific DSP systems – for this you will need to consult the appropriate user manual. Several very useful books have also been written which concentrate on practical applications of particular processors, e.g. El-Sharkawy (1996), Ingle and Proakis (1991) and Chassaing and Horning (1990).

4.21 CHAPTER SUMMARY

In this chapter we began with a very brief review of analogue filters, emphasizing that the frequency responses of real filters are usually a long way from that of ideal filters. We then looked at two general methods of designing IIR filters. The first was the 'direct' method. This entails placing poles and zeros in the z-plane so as to 'mould' the frequency response into the required shape. The approximate locations of the poles and zeros can be arrived at by calculation and then fine adjustments made to improve the response. Suitable CAD packages are invaluable when adopting this approach to filter design.

This direct method is fairly rough and ready and not feasible when a filter with a very 'tight' specification is required. We would then have to turn to the second approach, i.e. base the design on a suitable analogue filter. The analogue prototype, which will often be one of the 'classics', would then have to be converted to its digital equivalent using one of the various s-to-z transformations. Filter designs are usually based on lowpass filters as these can be readily converted to other types, both in the s-domain and the z-domain.

CAD packages are available which will carry out the complete design process.

4.22 PROBLEMS

1 A highpass filter is to be designed, using the 'direct' method, i.e. by placing transfer function poles and zeros at suitable points in the z-plane. The filter is to have a transfer function of the form $T(z) = k(z - 1)/(z - a)$, a cut-off frequency which is 0.125 of the sampling frequency ($0.25f_N$), and a gain of 1 at the Nyquist frequency.

2 Continuous filters with transfer functions of:

(a) $T(s) = \dfrac{s + 2}{s + 5}$

(b) $T(s) = \dfrac{15}{s^2 + 8s + 15}$

are to be converted to their digital equivalents. The bilinear transform is to be used, *with pre-warping*. Both continuous filters have breakpoint frequencies of 5 rad/s and this breakpoint is to be retained by the discrete filters, i.e. this is the frequency which must be pre-warped. Both filters are to be used with a sampling frequency of 8 Hz.

3 Convert the filter with the transfer function of $T(s) = s/(s + 4)$ to its discrete equivalent *using pole–zero mapping*. The filter is to be used with a sampling frequency of 4 Hz. The transfer function of the discrete filter must be scaled such that it has the same gain as the continuous filter at 2 Hz (f_N). (N.B. This is a *highpass* filter.)

4 The following transfer functions are for two lowpass, continuous filters. Use the *impulse-invariant* transformation to convert them to the transfer functions of equivalent discrete filters. The discrete filters must have the same d.c. gains as the corresponding analogue filter. The discrete filters are to be used with a sampling frequency of 5 Hz.

(a) $T(s) = \dfrac{8}{(s + 4)^2}$

(b) $T(s) = \dfrac{4(s + 3)}{(s + 2)(s + 6)}$

5 A lowpass analogue filter has a transfer function of $T(s) = 3/(s + 3)$. This transfer function is to be converted to one for a discrete, highpass filter, with a cut-off frequency of 6 rad/s. The filter is to be used with a sampling frequency of 10 Hz and the conversion is to be done by:

(a) changing from lowpass to highpass in the s-domain and then converting to the discrete, equivalent highpass filter by using the pole–zero mapping technique. (The discrete highpass filter must have the same gain as the continuous filter at the Nyquist frequency.)

(b) converting to an equivalent discrete, lowpass filter using the pole–zero mapping method (the added zero is to be placed at $z = -1$) and then converting from lowpass to highpass in the z-domain. (The continuous lowpass filter and its discrete equivalent must have the same d.c. gains.)

5   The design of FIR filters

5.1   CHAPTER PREVIEW

This, the final chapter, begins by making a case for the use of FIR filters rather than IIR filters *under certain conditions*. We will then look at two standard methods used to design FIR filters – the 'Fourier' or 'windowing' method and the 'frequency sampling' method. These two design techniques rely heavily on various forms of the Fourier transform and the inverse Fourier transform, and so these topics will also be covered, including the *'fast* Fourier transform' and its inverse.

5.2   INTRODUCTION

So far we have looked at various ways of designing IIR filters. Compared to FIR filters, IIR filters have the advantage that they need significantly fewer coefficients to produce a roughly equivalent filter. On the downside, they do have the disadvantage that they can be unstable if not designed properly, i.e. if any of the transfer function poles are outside the unit circle. However, with careful design this should not happen.

So why bother with FIR filters at all?

They have one major thing going for them which sometimes makes them preferable to IIR filters. This is that *they can be designed to have a linear phase response*.

5.3   PHASE-LINEARITY AND FIR FILTERS

Imagine a continuous signal, let's say a rectangular pulse, being passed through a filter. Also imagine that the transfer function of the filter is such that *the gain is 1 for all frequencies*. Such filters exist and are called 'all-pass' filters. It would seem reasonable that the pulse will pass through an all-pass filter undistorted. However, this is unlikely to be the case. This is because we have not taken into account the *phase response* of the filter. When the signal passes through a filter, the different frequencies making up the rectangular pulse will usually undergo different phase changes – effectively, signals of different frequencies are delayed by different times. It is as though, as a result of passing through the filter, signals are 'unravelled' and then put back together in a different way. This reconstruction results in distortion of the emerging signal.

   An example of a continuous, all-pass filter is one with the transfer function

$T(s) = (s - 4)/(s + 4)$. (Note that this transfer function has a zero on the right-hand side of the s-plane. This is fine – it is only poles which will cause instability if placed here.) This particular filter will have a gain of 1 for all frequencies – this should be fairly obvious from its p–z diagram. Interpretation of the signal response is made slightly easier if we imagine that we have an inverting amplifier in series with the filter. The frequency response, Fig. 5.1, confirms that the gain of the combination is 1. Figure 5.2 shows the response of the filter to a rectangular pulse – the signal has clearly been changed. Notice that the distortion occurs particularly at the leading and falling edges of the pulse. This would be expected, as the edges correspond to sudden changes in the signal magnitude and rapid changes consist of a broad band of signal frequencies. Figure 5.3 shows the output when a unit impulse passes through the filter. Theoretically, a unit impulse is composed of an infinite range of frequencies and so it should suffer major distortion – and it certainly does (notice how its width has spread to approximately 1 s). In a similar way, *discrete* all-pass filters will also cause distortion. Figure 5.4 shows what happens to a sampled rectangular pulse, consisting of three unit pulses, passed through such a filter.

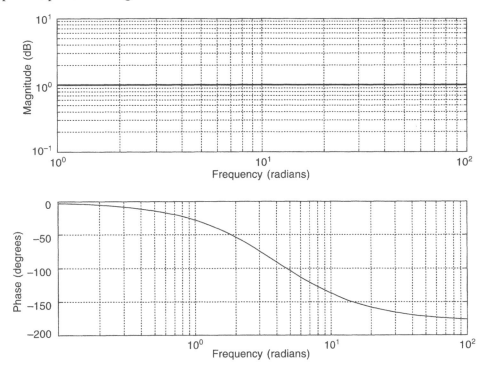

Figure 5.1

It can be shown that only if a filter is such that the gradient of the plot of phase shift against frequency is constant will there be no distortion of the signal due to the phase response.

This is because the effective signal delay introduced by a filter is given by $d\phi/d\omega$, where ϕ is the phase change. It follows that we require that $\phi = k\omega$, where k is a constant, if the delay is to be the same for all frequencies, as then $d\phi/d\omega = k$.

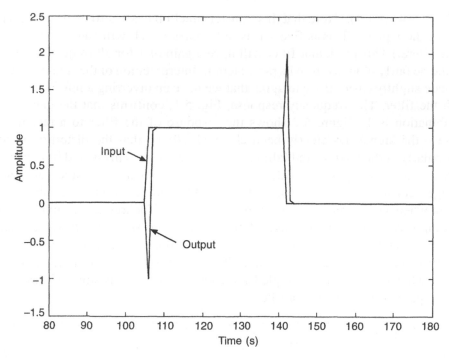

Figure 5.2

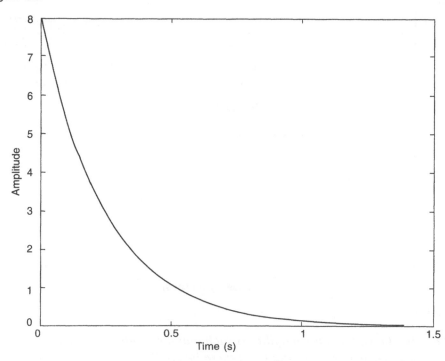

Figure 5.3

Clearly, the phase response of our all-pass filter is not linear (Fig. 5.1), and so the signals suffer distortion. If a filter has a phase response with a constant

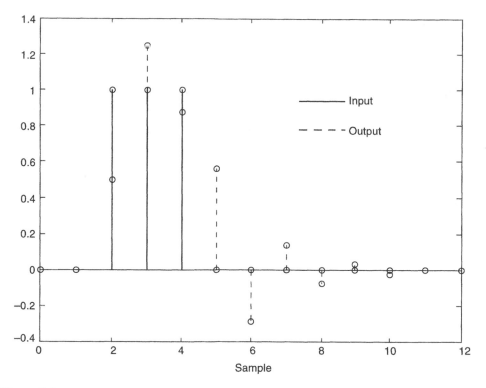

Figure 5.4

gradient, i.e. where the phase response is linear then, very sensibly, the filter is described as being a 'linear-phase filter'.

In the previous chapter we spent quite a lot of time converting continuous filters to their discrete IIR equivalents, and then comparing them very critically in terms of their magnitude responses. However, you might have noticed that not too much was made of any difference between their *phase responses*, even though the differences were sometimes very obvious. This is because the phase response of the original continuous filter itself was probably far from ideal and so it didn't matter *too* much if the phase response of the discrete filter differed from it.

While we're dealing with this subject, it should be mentioned that we can take an IIR filter with its poor, i.e. non-linear, phase response and place a suitable 'all-pass' discrete filter in series with it so as to linearize the combined phase responses. As well as having a gain of 1 for all signal frequencies, the compensating all-pass filter will have a phase response which is as close as possible to the inverse of that of the original filter.

So, to sum up, a case has been made for using FIR filters in preference to IIR filters, in certain circumstances. This is because FIR filters can be designed to have a linear phase response. As a result, such an FIR filter will not introduce any distortion into the output signal due to its phase characteristics. However, do not get the idea that all FIR filters are *automatically* linear-phase filters – far from it. If we want an FIR filter to have a linear phase response, and we usually do, then we need to *design* it to have this particular property.

5.4 RUNNING AVERAGE FILTERS

You met these filters way back in Chapter 1. Just to remind you, they are FIR filters that generate an output which is the average of the current input and a certain number of previous inputs. It is not a particularly useful filter practically – it behaves as a fairly ordinary lowpass filter. However, it is extremely useful to me at the moment in that *it is an example of an FIR filter that naturally has a linear phase characteristic*. For example, consider a running (or moving) average filter that averages the present and the previous three input samples, i.e.

$$Y(z) = \frac{X(z) + X(z)z^{-1} + X(z)z^{-2} + X(z)^{-3}}{4}$$

where $Y(z)$ and $X(z)$ are the z-transforms of the filter output and input respectively.

$$\therefore Y(z) = \frac{X(z)(1 + z^{-1} + z^{-2} + z^{-3})}{4} = \frac{X(z)(z^3 + z^2 + z + 1)}{4z^3}$$

or

$$T(z) = \frac{z^3 + z^2 + z + 1}{4z^3}$$

Figure 5.5 shows the frequency response for this filter. Note that the magnitude response is that of a lowpass filter with a cut-off frequency of about $0.2f_N$ but, much more importantly, *that the phase response is linear*. The discontinuity at $0.5f_N$ doesn't matter – what *is* important is that the gradient of the phase plot has a constant value.

So, what is it about the transfer function of a running average filter that makes it a linear-phase filter?

The secret is the symmetry of the numerator coefficients. In this example all numerator coefficients are 1, and so definitely *symmetric about the mid-point*. However, *any* FIR filter with this central symmetry will exhibit a linear phase response. For example, if we had a filter with the transfer function:

$$T(z) = 2 + z^{-1} + 3z^{-2} + z^{-3} + 2z^{-4}$$

or

$$T(z) = \frac{2z^4 + z^3 + 3z^2 + z + 2}{z^4}$$

i.e. symmetric coefficients of 2, 1, 3, 1, 2, then the phase response is as shown in Fig. 5.6 – i.e. once again a linear phase response, albeit split into segments.

So, to recap, when we design linear-phase FIR filters we must make sure that the coefficients of the transfer function numerator are symmetric about the centre, i.e. about the centre 'space' if there is an even number of coefficients and about the central coefficient if an odd number.

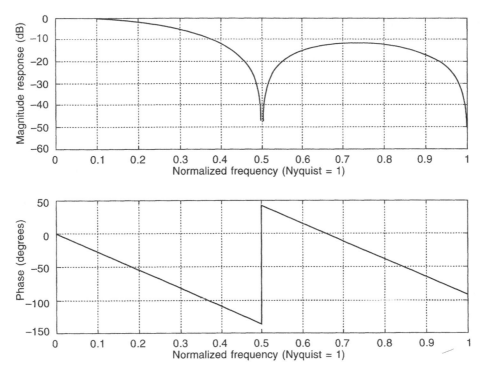

Figure 5.5

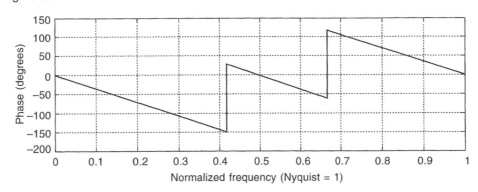

Figure 5.6

5.5 THE FOURIER TRANSFORM AND THE INVERSE FOURIER TRANSFORM

The first design method which we will use relies heavily on the Fourier transform, and so it makes sense to carry out a brief review of this extremely important topic at this point.

You are probably aware that the Fourier *series* allows us to express a periodic wave in terms of a d.c. value and a series of sinusoidal signals. The sinusoidal components have frequencies equal to the frequency of the original wave and harmonics of this 'fundamental' frequency. For example, a square wave consists of the fundamental frequency and weighted values of the *odd* harmonics. However, many signals are not periodic – speech would be an obvious example of a non-periodic waveform. This is where the *Fourier transform* is of such importance.

The Fourier transform, $F(j\omega)$, of a signal, $f(t)$, is defined by:

$$F(j\omega) = \int_{-\infty}^{\infty} f(t)e^{-j\omega t}\,dt$$

It is an extremely useful transform as it allows us to break a non-periodic signal, $f(t)$, down into its frequency components, $F(j\omega)$. $F(j\omega)$ is normally complex, expressing both the magnitude and the phase of the frequency components. As the signal is non-periodic the component frequencies will cover a continuous band, unlike the discrete frequencies composing a periodic wave.

The *inverse Fourier transform* (IFT) allows us to work the other way round (the clue is in the name!). In other words, if we use the IFT to operate on the *frequency spectrum* of a signal, we can recover the time variation, i.e. the signal shape. The IFT is given by:

$$f(t) = \frac{1}{2\pi} \int_{-\infty}^{\infty} F(j\omega)e^{j\omega t}\,d\omega \qquad (5.1)$$

As it is the *inverse* Fourier transform which is more relevant to us in the design of FIR filters, it's worth pausing here to look at an example of its use.

Let's imagine that a signal has the spectrum of Fig. 5.7, i.e. it consists of frequencies in the range of $-f_c$ to $+f_c$, all with equal weighting. As the Fourier transform of a signal is normally complex, it must be expressed in the form of an amplitude *and* a phase spectrum. For this particular signal we are assuming that the phase angle is zero for all frequencies, i.e. that its Fourier transform has no imaginary component. In other words, $F(j\omega) = 1$ for $-f_c \leq f \leq f_c$, else $= 0$.

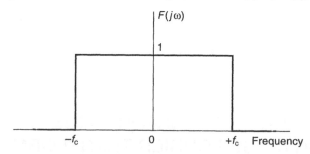

Figure 5.7

N.B. You might not be familiar with the concept of 'negative' frequencies. Clearly, a negative frequency is meaningless practically. However, representing signals in terms of their 'normal' positive frequencies and also the equivalent negative frequencies is a neat mathematical ploy which helps with the analysis of the signals.

Applying the inverse Fourier transform (equation (5.1)) with $F(j\omega) = 1$:

$$f(t) = \frac{1}{2\pi} \int_{-\omega_c}^{\omega_c} e^{j\omega t}\,d\omega \quad \text{where} \quad \omega_c = 2\pi f_c$$

N.B. We are now only integrating between $-\omega_c$ and $+\omega_c$ as $F(j\omega)$ has a value of zero outside this frequency range

$$\therefore f(t) = \frac{1}{2\pi jt} \left| e^{j\omega t} \right|^{\omega_c}_{-\omega_c} = \frac{1}{2\pi jt} (e^{j\omega_c t} - e^{-j\omega_c t})$$

but, from Euler's identity,

$$\sin \theta = \frac{1}{2j} (e^{j\theta} - e^{-j\theta})$$

$$\therefore f(t) = \frac{\sin \omega_c t}{\pi t} \tag{5.2}$$

This function is plotted in Fig. 5.8, using $\omega_c = 5$ rad/s and $t = -10$ s to $+10$ s. (There is nothing special about these values, they were chosen quite randomly.)

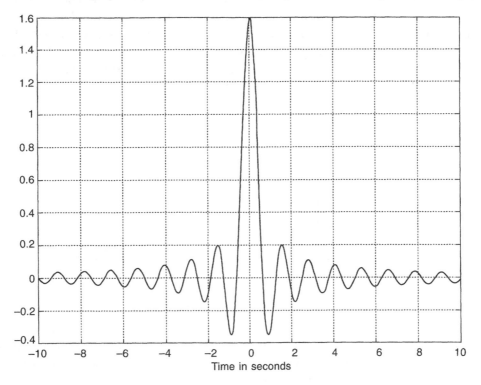

Figure 5.8

This signal might appear to be a little unusual and an unlikely one to meet on a regular basis! It is, in fact, an extremely important signal shape in the world of signal processing. Also, by using this simple frequency spectrum, the analysis was reasonably straightforward. You will find that this was a very useful example to have worked through when we move on to the design of an FIR in the next section.

So, to summarize, if we know the frequency spectrum of a signal we can use the IFT to derive the signal shape in the time domain. If we need to move in the opposite direction, i.e. from time domain to frequency domain, then we must use the Fourier transform.

This review is probably sufficient to allow us to move on but, if you feel that you need to do some more reading, there are plenty of circuit analysis/signal processing texts that deal with this topic. Some examples are: Howatson (1996), Denbigh (1998), Lynn and Fuerst (1994), Proakis and Manolakis (1996) and Meade and Dillon (1991). (*Keywords:* Fourier series, Fourier transform, inverse Fourier transform, amplitude spectrum, phase spectrum.)

5.6　THE DESIGN OF FIR FILTERS USING THE FOURIER TRANSFORM OR 'WINDOWING' METHOD

The starting point for this filter design method is the desired frequency response of the filter. To make use of the work done in the previous section, let's suppose that we want to design an FIR lowpass filter with a cut-off frequency of 4 kHz. The required magnitude response is shown in Fig. 5.9. We will decide on the sampling frequency during the design process. We will assume that the *phase response* is zero, i.e. that signals of all frequencies experience no phase change on passing through the filter.

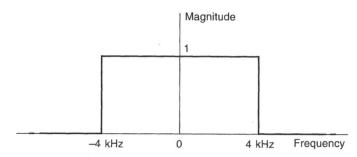

Figure 5.9

The first step in the design process is to *move from the frequency domain to the time domain, by using the inverse Fourier transform.* Luckily, we've already made this particular frequency-to-time conversion in the previous section (although for a frequency spectrum rather than a frequency response) and know that the corresponding time function, $f(t)$, is given by:

$$f(t) = \frac{\sin(\omega_c t)}{\pi t}$$

Therefore, for this filter, with its cut-off frequency of 4 kHz:

$$f(t) = \frac{\sin(8000\,\pi t)}{\pi t} \tag{5.3}$$

But what exactly does this time function represent?

It is the unit impulse response of the filter. To convince yourself of this, you need to imagine using a unit impulse as the filter input. As the Laplace transform of the unit impulse is '1', then the Laplace transform, $Y(s)$, of the output, will be the same as the filter transfer function, i.e. $Y(s) = 1 \times T(s) = T(s)$. As we are interested in the *frequency response*, we can replace 's' with $j\omega$. Therefore $Y(j\omega) = T(j\omega)$, where $Y(j\omega)$ is the frequency spectrum of the unit impulse response. We know that operating on the frequency spectrum of a signal with the IFT reconstructs the signal in the time domain. It follows that applying the IFT to $T(j\omega)$, and therefore to $Y(j\omega)$, must give us the time response of $Y(j\omega)$, which we know is the *unit impulse response of the filter*. Make sure that you are happy with this argument, as it is fundamental to the design process.

Figure 5.10(a) shows the plot of $f(t) = [\sin{(8000\pi t)}]/\pi t$, i.e. the filter unit impulse response, within the time window of -1.5 ms to $+1.5$ ms.

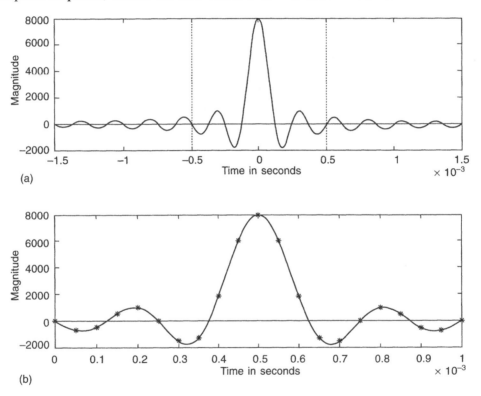

(a)

(b)

Figure 5.10

N.B. The time response shows a signal occurring before $t = 0$, i.e. before the input arrives, but don't worry about this for the moment.

The next step in the design is to sample the unit impulse response. *This is the crux of the design as these sampled values must be the FIR filter coefficients.* For example, let's imagine that we could represent the signal adequately by taking just five samples, *a, b, c, b, a*, with the samples being symmetric about the central peak. This, of course, is completely unrealistic and many samples would have to be taken – but bear with me.

If we now use these as the coefficients of the filter transfer function then this gives:

$$T(z) = a + bz^{-1} + cz^{-2} + bz^{-3} + az^{-4}$$

If we now use a unit sample sequence as the input to the discrete filter then the output will be the sequence a, b, c, b, a. It follows that the envelope of the unit sample response of the discrete filter will be the same as the unit impulse response of the continuous filter. In other words, *we will have designed a digital filter that has the unit sample response which corresponds to the required frequency response.*

This sounds fine, but we have a bit of a practical problem here in that the time response, theoretically, continues forever. Figure 5.10(a) shows only part of it; the ripples either side of the central peak gradually get smaller and smaller. Clearly we cannot take an infinite number of samples! What we have to do is compromise and sample over only a limited, but adequate, time span. Here we will sample between approximately $\pm 5 \times 10^{-4}$ s. This time slot is rather short but it will be adequate to demonstrate the principles of the approach. If we were to use a more realistic time window then the maths involved would become unnecessarily tedious and get in the way of the explanation.

If we wish one of our samples to be at the central peak, which seems a sensible thing to do, then, *as the coefficients must be symmetric about the centre to ensure phase-linearity*, it means that we will need to take an odd number of samples. Conveniently, if we sample every 0.5×10^{-4} s then we can fit exactly 21 samples into the 10^{-3} s time slot. We will therefore sample at 20 kHz ($T = 0.5 \times 10^{-4}$ s). This gives a Nyquist frequency of 10 kHz, which is well above the cut-off frequency of 4 kHz and so should be satisfactory. (Clearly, it would be pointless deciding to sample at a frequency that corresponds to a Nyquist frequency which is below the cut-off frequency!)

Before we carry out the sampling there is one other problem to address – that of the time response starting at a negative time. This has been referred to earlier but was put to one side. Clearly, this is an impossible response for a real digital filter as it means that a signal is generated before an input is received, in other words, the filter is 'non-causal'. However, this not a major problem *mathematically* as we can just shift the required time response to the right by 0.5×10^{-3} s so that it begins at $t = 0$. *Practically*, this has the effect of introducing a delay into the filter of 0.5×10^{-3} s, but this should not be a problem. It can be shown that it also changes the phase response of the filter from the original zero to a non-zero *but linear* phase response. Again, this is fine.

To reflect this time-shifting by 5×10^{-4} s, equation (5.3) needs to be changed to:

$$f(t) = \frac{\sin(8000\,\pi(t - 5 \times 10^{-4}))}{\pi(t - 5 \times 10^{-4})} \tag{5.4}$$

We are now, finally, in a position to sample the signal to obtain our filter coefficients. To do this we need to replace t with nT in equation (5.4), where n is the sample number, 0 to 20, and T the sampling period of 0.5×10^{-4} s.

$$\therefore f(nT) = \frac{\sin(8000\,\pi(nT - 5 \times 10^{-4}))}{\pi(nT - 5 \times 10^{-4})}$$

The 'time-windowed', shifted and sampled signal is shown in Fig. 5.10(b), and the sampled values in Table 5.1.

Table 5.1

n	$f(nT)$
0/20	0
1/19	−673
2/18	−468
3/17	535
4/16	1009
5/15	0
6/14	−1514
7/13	−1247
8/12	1871
9/11	6054
10	8000

N.B. The central value involves dividing 0 by 0, which is problematic! However, if we expand the sine function using the power series,

$$\sin x = x - \frac{x^3}{3!} + \frac{x^5}{5!} - \frac{x^7}{7!} + \dots$$

then we can see that $\sin x$ approaches x, as x approaches zero, and so $(\sin x)/x$ must approach 1. It follows that

$$\frac{\sin(8000\,\pi(nT - 5 \times 10^{-4}))}{\pi(nT - 5 \times 10^{-4})}$$

must become 8000 when $nT = 5 \times 10^{-4}$.

From Table 5.1, the filter must have the transfer function of:

$$T(s) = 0 - 673z^{-1} - 468z^{-2} + 535z^{-3} + 1009z^{-4} - 1514z^{-6} - 1247z^{-7} + 1871z^{-8}$$
$$+ 6054z^{-9} + 8000z^{-10} + 6054z^{-11} + 1871z^{-12} - 1247z^{-13} - 1514z^{-14}$$
$$+ 1009z^{-16} + 535z^{-17} - 468z^{-18} - 673z^{-19} + 0z^{-20}$$

or:

$$T(z) \frac{0z^{20} - 673z^{19} - 468z^{18} + 535z^{17} + \dots + 8000z^{-10} + \dots + 535z^3 - 468z^2 - 673z + 0}{z^{20}}$$

(The two zero coefficients have been included for clarity.)

So, now comes the moment of truth – the frequency response of the filter. This is shown in Fig. 5.11 – and it doesn't look too good! Although the filter has a linear phase response and the cut-off frequency is very close to the required 4 kHz ($0.4f_N$), the magnitude of the response is *far* too big. However, should we be too surprised by this? Think back to the previous chapter where we converted continuous filters to their discrete equivalents, using the impulse-invariant transform (Section 4.10). This method gave excellent agreement between the unit sample response of the discrete filter and the unit impulse response of its continuous prototype. However, although the magnitude response also had the correct shape, it was far too big, and so we had to scale the transfer function. This is exactly what has to be done here. The magnitude of the response in the passband is approximately 86 dB rather than the 0 dB required, which is approximately 20 000 times too big, and so we need to divide all coefficients by 20 000.

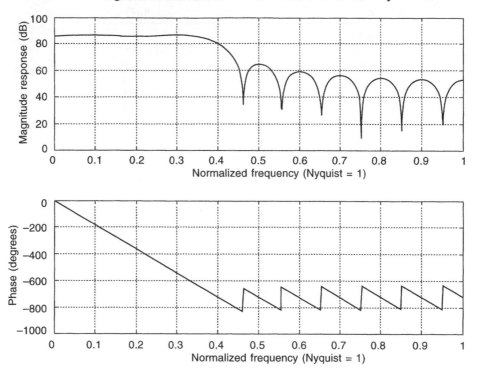

Figure 5.11

The magnitude response of the scaled filter is shown in Fig. 5.12 and now agrees closely with the filter specification. It follows that our filter has the transfer function of:

$$T(z) = \frac{0z^{20} - 673z^{19} - 468z^{18} + 535z^{17} \ldots 535z^{3} - 468z^{2} - 673z + 0}{20\,000\,z^{20}}$$

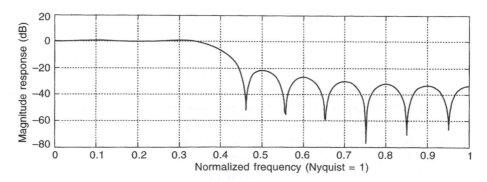

Figure 5.12

N.B. It is no coincidence that we are sampling at 20 kHz and that we needed to scale by 1/20 000. It is possible to show that, to derive the filter coefficients from the sampled unit impulse response values, *we need to divide the sampled values by the sampling frequency.*

So, to summarize: the starting point in the design process is the required frequency response of the filter. The IFT is then taken of this frequency response to generate the filter unit impulse response. An adequate time window is now chosen, with the central peak of the unit impulse response at its mid-point. The selected section of the unit impulse response is then shifted, sampled symmetrically, and the samples scaled. These scaled values are the filter coefficients.

A quicker approach

To make the principles behind the design process as clear as possible, the filter design has been approached in the most logical way. We started with the required frequency response and then used the IFT to convert to the time domain to get the corresponding unit impulse response. The unit impulse response was then sampled, shifted and finally scaled to find the FIR filter coefficients. However, this is rather a lengthy process and a much more direct and practical way of finding the coefficients follows.

We know that the unit impulse response is given by $f(t) = (\sin(2\pi f_c t))/\pi t$ from equation (5.2), and so the sampled values, *before shifting and scaling*, are given by:

$$f(nT) = \frac{\sin(2\pi f_c nT)}{\pi nT}$$

However, we know that to get the filter coefficients, $C(n)$, we need to scale these samples by dividing by f_s.

$$\therefore C(n) = \frac{\sin(2\pi f_c nT)}{f_s \pi nT}$$

But $f_s T = 1$,

$$\therefore C(n) = \frac{\sin(2\pi f_c nT)}{\pi n}$$

It is often more convenient to express the cut-off frequency in the normalized form, i.e. as a fraction of the Nyquist frequency, f_N, and so we can rewrite the equation for $C(n)$ as:

$$C(n) = \frac{\sin\left(2\pi \dfrac{f_c}{f_N} f_N nT\right)}{\pi n}$$

But $2f_N T = f_s T = 1$,

$$\boxed{\therefore C(n) = \frac{\sin\left(n\pi \dfrac{f_c}{f_N}\right)}{n\pi}}$$

By making *this* our starting point and, for our particular example, using $n = -10$ to $+10$, we can find the filter coefficients very quickly. We do not have to bother to time-shift the expression as long as we remember that $C(-10)$ will correspond to our first coefficient, $C(-9)$ to the second, etc. Neither is scaling necessary as this has been incorporated into the expression for $C(n)$.

5.7 WINDOWING AND THE GIBBS PHENOMENON

Although the filter we have just designed satisfies the specification quite well, you will have noticed the 'ripples' in the stopband (Fig. 5.12). Much more seriously, these are also present in the passband, although they are less obvious because of the log scale. Figure 5.13 shows an expanded version of just the passband – the ripple is now much more apparent. This ripple in the passband is obviously undesirable as, ideally, we require a constant gain over this frequency range.

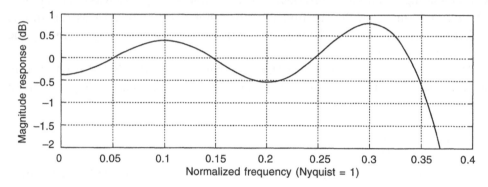

Figure 5.13

This rippling is termed the 'Gibbs' phenomenon, after an early worker in the field of signal analysis. It occurs because we have applied a 'window' to the unit impulse response, i.e. we only considered that portion of the signal contained within the time-slot of 10^{-3} s.

In the previous example we have effectively applied a *rectangular* window, in that all signal values were multiplied by 1 over the 10^{-3} s time slot and by 0 for all other time values. There are, however, other standard window profiles available to us *which we can use to reduce the ripple*. One that is widely used is the *Hamming* window. The profile of this particular window is defined by:

$$w(n) = 0.54 - 0.46 \cos\left(\frac{2n\pi}{N-1}\right)$$

where n is the sample number and N the total number of samples. The Hamming window profile, for $N = 21$, is shown in Fig. 5.14.

What we have to do is multiply all of the sampled values within the 10^{-3} s time slot by the corresponding window value, before we find our filter coefficients. More conveniently, this is equivalent to just multiplying each of our 'raw' filter coefficients by the Hamming window values.

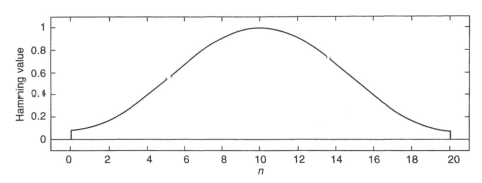

Figure 5.14

The Hamming window values and the new filter coefficients for our lowpass filter are given in Table 5.2.

Table 5.2

n	'Old' coefficients	Window values	New coefficients
0/20	0	0.080	0
1/19	-3.37×10^{-2}	0.103	-3.45×10^{-3}
2/18	-2.34×10^{-2}	0.168	-3.93×10^{-3}
3/17	2.68×10^{-2}	0.270	7.21×10^{-3}
4/16	5.05×10^{-2}	0.398	2.01×10^{-2}
5/15	0	0.54	0
6/14	-7.57×10^{-2}	0.682	-5.16×10^{-2}
7/13	-6.24×10^{-2}	0.810	-5.05×10^{-2}
8/12	9.36×10^{-2}	0.912	8.53×10^{-2}
9/11	3.03×10^{-2}	0.978	2.96×10^{-1}
10	0.4	1	0.4

The response for the new filter is shown in Fig. 5.15. Applying the Hamming window has clearly been a success as the ripple in the passband is *greatly* reduced compared to Fig. 5.13.

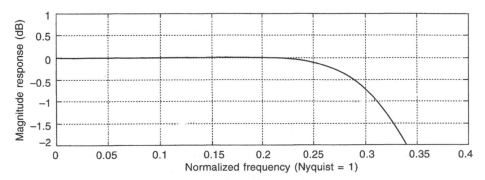

Figure 5.15

N.B.1   There are several other window profiles which are commonly used. For example:

$$\text{Hanning: } w(n) = \frac{1}{2}\left[1 - \cos\left(\frac{2\pi n}{N-1}\right)\right]$$

$$\text{Blackman: } w(n) = 0.42 - 0.5 \cos\left(\frac{2\pi n}{N-1}\right) + 0.08 \cos\left(\frac{4\pi n}{N-1}\right)$$

N.B.2   The MATLAB function 'fir1' will generate FIR filter transfer function coefficients – the Hamming window function is automatically included. For example, all that is needed to design our lowpass filter is the single instruction:

$a = \text{fir1}(20,0.4)$

where the '20' is the number of coefficients less 1, and the '0.4' is the normalized cut-off frequency, i.e. f_c/f_N.

Other window functions can be used. Refer to the MATLAB user manual or help screen for further details.

5.8   HIGHPASS, BANDPASS AND BANDSTOP FILTERS

We have shown that the coefficients for a linear-phase *lowpass* FIR filter are given by:

$$C(n) = \frac{\sin\left(n\pi \dfrac{f_c}{f_N}\right)}{n\pi}$$

But what if we need a highpass filter?

In Chapter 4, a lowpass IIR filter was converted to a highpass IIR filter by applying a particular transform in the z-domain (Section 4.17). We *could* do this here but it would be an unenviable task as there are so many z terms to transform. Also, even if we bothered, we would end up with an *IIR* filter, and one with a non-linear phase response. However, all is not lost, as it can be shown that the unit sample responses, *and so the filter coefficients*, for linear-phase highpass, bandpass and bandstop FIR filters are as shown in Table 5.3, where f_u is the upper cut-off frequency and f_1 the lower. If you require more details of the origin of these expressions see Chassaing and Horning (1990).

Highpass, bandpass and bandstop filters, with a variety of window functions, can also be designed using the MATLAB, 'fir1' function. Refer to the MATLAB user manual or help screen for further details.

5.9   SELF-ASSESSMENT TEST

1 (a) Design a lowpass FIR filter with a cut-off frequency of 3 kHz. The filter is to be used with a sampling frequency of 12 kHz and is to have a 'length' of 11, i.e. 11 coefficients are to be used.

Table 5.3

Filter type	$C(n)$ for $n \neq 0$	$C(n)$ for $n = 0$
Highpass	$-\dfrac{\sin\left(n\pi \dfrac{f_c}{f_N}\right)}{n\pi}$	$1 - \left(\dfrac{f_c}{f_N}\right)$
Bandpass	$\dfrac{\sin\left(n\pi \dfrac{f_u}{f_N}\right)}{n\pi} - \dfrac{\sin\left(n\pi \dfrac{f_1}{f_N}\right)}{n\pi}$	$\left(\dfrac{f_u}{f_N}\right) - \left(\dfrac{f_1}{f_N}\right)$
Bandstop	$\dfrac{\sin\left(n\pi \dfrac{f_1}{f_N}\right)}{n\pi} - \dfrac{\sin\left(n\pi \dfrac{f_u}{f_N}\right)}{n\pi}$	$1 - \left(\dfrac{f_u}{f_N} - \dfrac{f_1}{f_N}\right)$

(b) Redesign the filter, so as to reduce the passband ripple, by using a Hanning (N.B. *not* Hamming!) window.

2 Design a *highpass* FIR filter with cut-off frequency of 4 kHz. The filter is to be used with a sampling frequency of 20 kHz. A rectangular window is to be applied and 21 samples are to be taken. (Hint: Most of the groundwork has already been done during the design of the lowpass filter in Section 5.6.)

5.10 RECAP

- To design an FIR lowpass filter, using the Fourier (or windowing) method, we need to find the filter coefficients, $C(n)$, by using

$$C(n) = \frac{\sin\left(n\pi \dfrac{f_c}{f_N}\right)}{n\pi}$$

  We are effectively sampling, shifting and scaling the unit impulse response of the desired filter.

- If necessary, in order to reduce passband (and stopband) ripple, i.e. the 'Gibbs phenomenon', a suitable window function (other than rectangular) can be applied.

- Expressions also exist for deriving the transfer function coefficients for highpass, bandpass and bandstop filters.

5.11 THE DISCRETE FOURIER TRANSFORM AND ITS INVERSE

The Fourier and inverse Fourier transforms were reviewed in Section 5.5. This was because the 'Fourier', or 'windowing', method of designing FIR filters relied heavily on the inverse Fourier transform. The other common design method is based on a variation of the inverse Fourier transform, termed the *discrete* inverse Fourier transform. A few pages on this transform are therefore in order.

Like the 'normal' or continuous Fourier transform, the discrete Fourier transform (DFT) allows us to find the frequency components that make up a signal. However, there is a major difference in that the DFT is an approximation as it converts a *sampled time signal* into the corresponding *sampled frequency spectrum*. In other words, the frequency spectrum derived from the sampled signal is defined only at particular frequency values. If the sampling frequency is f_s, and N signal samples are taken, then the frequency spectrum generated by the DFT will consist only of frequencies of kf_s/N, where $k = 0$ to $N - 1$. For example, if we sampled a signal at 5 kHz and we took just 10 samples, then the amplitude spectrum unravelled by the DFT might look like Fig. 5.16, i.e. only the amplitudes of the signal components at 0, 500, 1000, 1500, . . . , 4500 Hz will be revealed. The 'step' frequency of 500 Hz, (f_s/N), is called the *frequency resolution*.

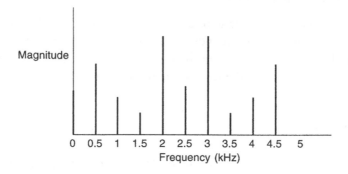

Figure 5.16

Also note the symmetry of the amplitude spectrum about the Nyquist frequency of 2.5 kHz. *All* discrete amplitude spectra, resulting from the spectral analysis of *sampled signals*, exhibit this symmetry. To understand the symmetry, imagine that the sequence is the output from a discrete filter, in response to an input of a unit sample function. As the z-transform of the output $Y(z) = 1 \times T(z)$, then the sampled spectrum of the signal must be the same as the sampled frequency response of the filter, i.e. the DFTs of the two must be the same. Now imagine the 'frequency point' moving in an anticlockwise direction around the unit circle of the p–z diagram for the filter transfer function. When the frequency point passes the Nyquist frequency the amplitude spectrum must begin to repeat itself, due to the symmetry of the p–z diagram about the real axis. For this reason, the Nyquist frequency is sometimes referred to as the 'folding' frequency.

Anyway – back to the main story.

Sampled frequency components are just what we need when we are processing frequency data by means of a computer, and so the DFT serves this purpose very well. However, a word of caution. If anything unusual is happening *between* the sampled frequencies then this will be missed when the DFT is applied. For example, let's say that there is a 'spike' in the amplitude spectrum at a frequency of 660 Hz. If we are sampling at 5 kHz, and taking 10 samples as before, then this spike will fall between the displayed, sampled frequencies of 500 Hz and 1 kHz and so will not be revealed. Clearly, the more samples taken, the better the frequency resolution and so the less that will be missed, i.e. the more closely the DFT will approximate to the 'normal' or continuous Fourier transform.

As might be expected, the *inverse discrete Fourier transform* (IDFT) performs the opposite process, in that it starts with the sampled frequency spectrum and reconstructs the sampled signal from it. It is the IDFT which will be of particular use to us when we move on to the next design method.

The discrete Fourier transform (sampled time to sampled frequency) is defined by:

$$X_k = \sum_{n=0}^{N-1} x[n]e^{-j\frac{2\pi nk}{N}}$$

(5.5)

where X_k is the kth value of the frequency samples.

The *inverse* discrete Fourier transform (sampled frequency to sampled time) is defined by:

$$x[n] = \frac{1}{N} \sum_{k=0}^{N-1} X_k e^{j\frac{2\pi nk}{N}}$$

(5.6)

where $x[n]$ is the nth sample in the sampled signal.

To make more sense of these two very important transforms it's best to look at an example.

Example 5.1

Use the DFT to find the sampled frequency spectrum, both magnitude and phase, of the following sequence: 2, −1, 3, given that the sampling frequency is 6 kHz. When the sampled spectrum has been derived, apply the inverse DFT to regain the sampled signal.

Solution

As we are sampling at 6 kHz, and taking three samples (very unrealistic but O.K. to use just to demonstrate the process), the frequency resolution is 6/3 (f_s/N), i.e. 2 kHz. Therefore the DFT will derive the sampled spectrum for frequency values of 0, 2 and 4 kHz (kf_s/N for $n = 0$ to $N - 1$, where $N = 3$).

We now need to apply the DFT to find the sampled frequency spectrum, *both magnitude and phase*. Therefore, using

$$X_k = \sum_{n=0}^{N-1} x[n]e^{-j\frac{2\pi nk}{N}} \quad \text{(for } N = 3\text{)}$$

- $k = 0$:

$$X_0 = \sum_{n=0}^{2} x[n]e^{-0} = \sum_{n=0}^{2} x[n] = 2 - 1 + 3 = 4\angle 0°$$

- $k = 1$:

$$X_1 = \sum_{n=0}^{2} x[n]e^{-j\frac{2\pi n}{N}} = 2e^{-j0} - 1e^{-j\frac{2\pi}{3}} + 3e^{-j\frac{2\pi 2}{3}}$$

$$\therefore X_1 = 2 - e^{-j\frac{2\pi}{3}} + 3e^{-j\frac{4\pi}{3}}$$

Using Euler's identity:

$$X_1 = 2 - (\cos 0.67\pi - j \sin 0.67\pi) + 3(\cos 1.33\pi - j \sin 1.33\pi)$$

$$\therefore X_1 = 2 + 0.51 + j0.86 - 1.53 + j2.58 = 0.98 + j3.44$$

$$\therefore X_1 = 3.58\angle 74.1°$$

- $k = 2$:

$$X_2 = \sum_{n=0}^{2} x[n]e^{-j\frac{2\pi n 2}{3}} = 2 - e^{-j\frac{4\pi}{3}} + 3^{-j\frac{8\pi}{3}}$$

$$\therefore X_2 = 2 - (\cos 1.33\pi - j \sin 1.33\pi) + 3(\cos 2.66\pi - j \sin 2.66\pi)$$

$$= 2 + 0.51 - j0.86 - 1.45 - j2.63 = 1.06 - j3.49$$

$$\therefore X_2 = 3.65\angle -73.1°$$

The magnitude and phase values are shown in Fig. 5.17.

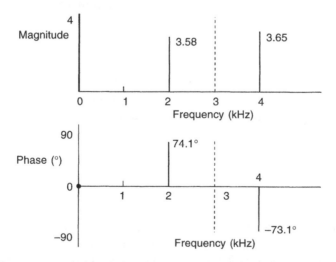

Figure 5.17

N.B.   As explained earlier in this section, the magnitudes at 2 and 4 kHz should be the same, i.e. *symmetric about the Nyquist frequency* of 3 kHz – the difference is due to approximations made during calculation. Similarly, the phase angles should have *inverted* symmetry, in other words, have the same magnitudes but with opposite signs.

We should now be able to process these three sampled frequency values of $4\angle 0$, $3.58\angle 74.1°$ and $3.65\angle -73.1°$ using the *IDFT*, to get back to the three signal samples of 2, –1 and 3.

Using

$$x[n] = \frac{1}{N} \sum_{k=0}^{N-1} X_k e^{j\frac{2\pi nk}{N}} \quad \text{(the IDFT)}$$

- $n = 0$:

$$x[0] = \frac{1}{3} \sum_{k=0}^{2} X_k e^{j0} = \frac{1}{3} [4 + (0.98 + j3.44) + (1.06 - j3.49)]$$

$$= 2.01 - j0.017$$

$$\therefore x[0] = 2$$

(Assuming that the relatively small imaginary part is due to approximations made during the maths, which seems a reasonable assumption. Note that this *is* the first signal sample.)

- $n = 1$:

$$x[1] = \frac{1}{3} \sum_{k=0}^{2} X_k e^{j\frac{2\pi k}{3}}$$

$$\therefore x[1] = \frac{1}{3} \left[4 + (0.98 + j3.44)e^{j\frac{2\pi}{3}} + (1.06 - j3.49)e^{j\frac{4\pi}{3}} \right]$$

$$= \frac{1}{3} \left[4 + (0.98 + j3.44)\left(\cos \frac{2\pi}{3} + j \sin \frac{2\pi}{3} \right) \right.$$

$$\left. + (1.06 - j3.49)\left(\cos \frac{4\pi}{3} + j \sin \frac{4\pi}{3} \right) \right]$$

Missing out the next few lines of complex maths (which you might wish to work through for yourself), and ignoring small approximation errors, we end up with $x[1] = -1$.

- $n = 2$:

$$x[2] = \frac{1}{3} \sum_{k=0}^{2} X_k e^{j\frac{4\pi k}{3}}$$

$$\therefore x[2] = \frac{1}{3} \left[4 + (0.98 + j3.44)e^{j\frac{4\pi}{3}} + (1.06 - j3.49)e^{j\frac{8\pi}{3}} \right]$$

$$= \frac{1}{3} \left[4 + (0.98 + j3.44)\left(\cos \frac{4\pi}{3} + j \sin \frac{4\pi}{3} \right) \right.$$

$$\left. + (1.06 - j3.49)\left(\cos \frac{8\pi}{3} + j \sin \frac{8\pi}{3} \right) \right]$$

Leaving out the intervening steps, and again ignoring small rounding errors, we get $x[2] = 3$, which agrees with the final signal sample.

5.12   THE DESIGN OF FIR FILTERS USING THE 'FREQUENCY SAMPLING' METHOD

The starting point for the previous 'Fourier' or 'windowing' design method (Section 5.6), was the required frequency response. The IFT was then applied to convert the frequency response to the unit impulse response of the filter. The unit impulse response was then shifted, sampled and scaled to give the filter coefficients.

The 'frequency sampling' method is very similar but the big difference here is that the starting point is the *sampled* frequency response. This can then be converted *directly* to the FIR filter unit sample response, and so to the filter coefficients, by using the *IDFT*.

To demonstrate this approach we will design a lowpass FIR filter with a cut-off frequency of 2 kHz – the filter being used with a sampling frequency of 9 kHz. Therefore $f_c = 0.444f_N$.

The required magnitude response of the filter is shown in Fig. 5.18. For the purpose of demonstrating this technique the frequency has been sampled at just

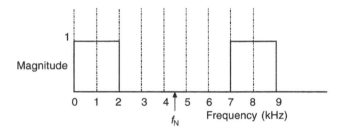

Figure 5.18

nine points, starting at 0 Hz, i.e. every 1 kHz. Note that f_s is *not* one of the sampled frequencies. To understand this, look back to Example 5.1, where a sampled time signal was converted to its corresponding sampled frequency spectrum using the DFT. By doing this, only frequencies up to $(N-1)f_s/N$ were generated, i.e. the sampled frequency spectrum stopped one sample away from f_s. As would be expected, by using the IDFT to process these frequency samples, the original signal was reconstituted. With the frequency sampling method, we start with the sampled frequency response, *stopping one sample short of f_s*, and aim to derive the equivalent sampled time function by applying the IDFT. As explained in Section 5.6, this time function will be the filter unit sample response, i.e. $Y(z) = 1 \times T(z)$, and so the IDFT of $T(z)$ must be the IDFT for $Y(z)$. These sampled signal values must also be the FIR filter coefficients.

The frequency samples for the required filter, derived from Fig. 5.18, are shown in Table 5.4.

Two important points to note. The first is that the response samples shown in Table 5.4 are the *magnitudes* of the sampled frequency response, $|X_k|$, whereas the X_k in the expression for the IDFT, equation (5.6), *also includes the phase angle*.

The second point is that the samples at $n = 2$ and $n = 7$ are shown as 0.5 rather

Table 5.4

n	$\mid X_k \mid$
0	1
1	1
2	0.5
3	0
4	0
5	0
6	0
7	0.5
8	1

than 1 or 0. By using the average of the two extreme values, we will design a much better filter, i.e. one that satisfies the specification more closely than if either 1 or 0 had been used.

We will now insert our frequency samples into the IDFT in order to find the sampled values of the unit impulse response and, hence, the filter coefficients. However, before we do this, a few changes will be made to the IDFT, so as to make it more usable.

At the moment our IDFT is expressed in terms of the complex values X_k, rather than the magnitude values, $\mid X_k \mid$, of Table 5.4. However, it can be shown that $\mid X_k \mid$ and X_k are related by:

$$\boxed{X_k = \mid X_k \mid e^{-j\frac{2\pi k}{N}\left(\frac{N-1}{2}\right)}}$$

or $X_k = \mid X_k \mid e^{-jk\pi\left(\frac{8}{9}\right)}$ for this particular example.

N.B. This exponential format is just a way of including the phase angle, i.e.

$$X_k = \mid X_k \mid e^{-j\frac{2\pi k}{N}\left(\frac{N-1}{2}\right)} = \mid X_k \mid \left(\cos\frac{2\pi k}{N}\left(\frac{N-1}{2}\right) - j\sin\frac{2\pi k}{N}\left(\frac{N-1}{2}\right)\right)$$

(from Euler's identity).

$$\therefore \tan\phi = -\frac{\sin\dfrac{2\pi k}{N}\left(\dfrac{N-1}{2}\right)}{\cos\dfrac{2\pi k}{N}\left(\dfrac{N-1}{2}\right)} = -\tan\frac{2\pi k}{N}\left(\frac{N-1}{2}\right)$$

where ϕ is the phase angle.

Therefore the *magnitude* of the phase angle is

$$\frac{2\pi k}{N}\left(\frac{N-1}{2}\right)$$

Although it's not obvious, this particular phase angle indicates a filter delay of $(N-1)/2$ sample periods. When we designed FIR filters using the 'Fourier' method in Section 5.6, we assumed zero phase before shifting, and then shifted our time response by $(N-1)/2$ samples to the right, i.e. we effectively introduced a filter delay of $(N-1)/2$ samples. This time-shifting results in a filter which has a phase response which is no longer zero for all frequencies *but the new phase response is linear*. In other words, by using this particular expression for the phase angle, we are imposing an appropriate constant filter delay for all signal frequencies. The exponential expression of $e^{-j\frac{2\pi k}{N}\left(\frac{N-1}{2}\right)}$ is the same for all linear-phase FIRs that have a delay of $(N-1)/2$ samples – which is the norm. For a fuller explanation, see Damper (1995).

Therefore, replacing X_k with $|X_k|e^{-j\frac{2\pi k\alpha}{N}}$, in the IDFT (equation (5.6)), where $\alpha = (N-1)/2$:

$$x[n] = \frac{1}{N}\sum_{k=0}^{N-1}|X_k|e^{-j\frac{2\pi k\alpha}{N}}e^{j\frac{2\pi kn}{N}} = \frac{1}{N}\sum_{k=0}^{N-1}|X_k|e^{j\frac{2\pi k(n-\alpha)}{N}}$$

Therefore, from Euler's identity:

$$x[n] = \frac{1}{N}\sum_{k=0}^{N-1}|X_k|(\cos[2\pi k(n-\alpha)/N] + j\sin[2\pi k(n-\alpha)/N])$$

Further, as $x[n]$ values must be real, i.e. they are actual signal samples, then we can ignore the imaginary 'sin' terms.

$$\therefore x[n] = \frac{1}{N}\sum_{k=0}^{N-1}|X_k|\cos[2\pi k(n-\alpha)/N]$$

Finally, as all X terms, apart from X_0, are partnered by an equivalent symmetric sample, i.e. $|X_1| = |X_8|$, $|X_2| = |X_7|$ etc. (Fig. 5.18), then we can further simplify to:

$$x[n] = \frac{1}{N}\left[\sum_{k=1}^{(N-1)/2}2|X_k|\cos[2\pi k(n-\alpha)/N] + X_0\right] \tag{5.7}$$

We now only have to sum between $k = 1$ and $k = (N-1)/2$ rather than 0 and $(N-1)$ and so avoid some work! In our example this means summing between 1 and 4 rather than 0 and 8.

To recap, we can use equation (5.7) to find the values, $x[n]$, of the filter unit sample response. *These values are also the FIR filter coefficients.*

O.K – so let's use equation (5.7) to find the filter coefficients. Substituting $n = 0$ and $\alpha = (9-1)/2 = 4$ into equation (5.7), and carrying out the summation:

$$x[0] = \frac{1}{9}[2(\cos[2\pi(0-4)/9] + 0.5\cos[4\pi(0-4)/9]) + 1] = -0.0126$$

As $|X_3|$ and $|X_4| = 0$ we only have two terms to sum.

- $n = 1$:

$$x[1] = \frac{1}{9}[2(\cos[2\pi(1-4)/9] + 0.5\cos[4\pi(1-4)/9]) + 1] = -0.0556$$

- $n = 2$:

$$x[2] = \frac{1}{9}[2(\cos[2\pi(2-4)/9] + 0.5\cos[4\pi(2-4)/9]) + 1] = 0.0453$$

Check for yourself that $x[3] = 0.3006$ and $x[4] = 0.4444$.

N.B. As we are designing a *linear-phase* filter, then the signal samples must be symmetric about the central value, $x[4]$, i.e. $x[0] = x[8]$, $x[1] = x[7]$, etc., and so we only need to find the five $x[n]$ values, $x[0]$ to $x[4]$.

It follows that the nine samples of the filter unit sample response are:

−0.0126, −0.0556, 0.0453, 0.3006, 0.444, 0.3006, 0.0453, −0.0556, −0.0126

and, as these must also be our filter coefficients:

$$T(z) = -0.0126 - 0.0556z^{-1} + 0.0453z^{-2} + 0.3006z^{-3} + 0.444z^{-4} + 0.3006z^{-5} \\ + 0.0453z^{-6} - 0.0556z^{-7} - 0.0126z^{-8}$$

or

$$T(z) = \frac{-0.0126z^{8} - 0.0556z^{7} + \ldots + 0.444z^{4} + \ldots - 0.0556z - 0.0126}{z^{8}}$$

Figure 5.19 shows the frequency response of our filter and it looks pretty good. The cut-off frequency is not exactly the 2 kHz ($0.444f_N$) required, but it's not too far away. To be fair, to save too much mathematical processing getting in the way of the solution, far too few frequency samples were used. If we had taken a more realistic number of frequency samples than nine then the response would obviously have been much better.

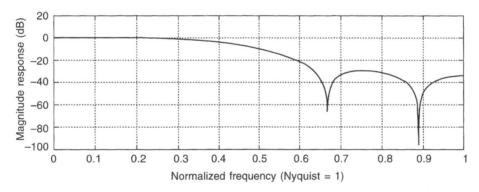

Figure 5.19

Here we used an odd number of frequency samples. The expression for the IDFT for an *even* number of samples is given in equation (5.8a). For completeness, the expression for an odd number is repeated as equation (5.8b).

$$x[n] = \frac{1}{N}\left[\sum_{k=1}^{(N/2)-1} 2\,|X_k|\cos\left[2\pi k(n-\alpha)/N\right] + X_0\right] \quad \text{(for even } N\text{)} \quad (5.8a)$$

$$x[n] = \frac{1}{N}\left[\sum_{k=1}^{(N-1)/2} 2\,|X_k|\cos\left[2\pi k(n-\alpha)/N\right] + X_0\right] \quad \text{(for odd } N\text{)} \quad (5.8b)$$

The 'even' expression is the same as for an odd number of samples apart from the k range being from 1 to $(N/2) - 1$ rather than 1 to $(N - 1)/2$. If you inspect equation (5.8a) you will notice that the $N/2$ sample is not included in the IDFT. For example, if we used eight samples, then X_0, X_1, X_2 and X_3 only would be used, and not X_4. I won't go into this in any detail but using an even number of samples does cause a slight problem in terms of the $N/2$ frequency term. One solution, which gives acceptable results, is to treat $X_{N/2}$ as zero – this is effectively what has been done in equation (5.8a). See Damper (1995) for an alternative approach.

5.13   SELF-ASSESSMENT TEST

A *highpass* FIR filter is to be designed, using the 'frequency sampling' technique. The filter is to have a cut-off frequency of 3 kHz and is to be used with a sampling frequency of 9 kHz. The frequency response is to be sampled every 1 kHz.

5.14   RECAP

- The discrete Fourier transform (DFT), is used to convert a sampled signal to its sampled frequency spectrum, while the inverse discrete Fourier transform (IDFT) achieves the reverse process.

- The sampled frequency spectrum obtained using the DFT consists of frequency components at frequencies of kf_s/N, for $k = 0$ to $N - 1$, where f_s is the sampling frequency and N the number of samples taken.

- By beginning with the desired sampled frequency response of a filter, it is possible to find the coefficients of a suitable linear-phase FIR filter by applying the IDFT.

5.15   THE FAST FOURIER TRANSFORM AND ITS INVERSE

Clearly, the discrete Fourier transform is hugely important in the field of signal processing as it allows us to analyse the spectral content of a sampled signal. The IDFT does the opposite, and we have seen how invaluable it is when it comes to the design of FIR filters. Although we will not be looking at this particular method, the DFT can *also* be used to design digital filters.

However, as useful as they are, the DFT and its inverse have a major practical problem – this is that they involve a great deal of computation and so are very extravagant in terms of computing time. Earlier, we performed some transforms manually, using unrealistically small numbers of samples. Imagine the computation needed if thousands of samples were to be used – and this is not unusual in practical digital signal processing. Because of this, algorithms have been developed which allow the two transforms to be carried out much more quickly – very sensibly they are called the *fast Fourier transform* (FFT), and the *fast inverse Fourier transform* (FIFT) or *inverse fast Fourier transform* (IFFT). Basically, these algorithms break down the task of transforming sampled time to sampled frequency, and vice-versa, into much smaller blocks – they also take advantage of any computational redundancy in the 'normal' discrete transforms.

To demonstrate the general principles, the FFT for a sequence of four samples, $x[0]$, $x[1]$, $x[2]$ and $x[3]$, i.e. a 'four point' sequence, will now be derived.

From

$$X_k = \sum_{n=0}^{N-1} x[n]e^{-j\frac{2\pi nk}{N}} \quad \text{(the DFT)}$$

- $k = 0$:

$$X_0 = x[0] + x[1] + x[2] + x[3]$$

- $k = 1$:

$$X_1 = x[0] + x[1]e^{\frac{-j2\pi}{4}} + x[2]e^{\frac{-j4\pi}{4}} + x[3]e^{\frac{-j6\pi}{4}}$$

or, more neatly:

$$X_1 = x[0] + x[1]W + x[2]W^2 + x[3]W^3$$

where $W = e^{\frac{-j2\pi}{4}}$. (W is often referred to as the 'twiddle factor'!)

- $k = 2$:

$$X_2 = x[0] + x[1]W^2 + x[2]W^4 + x[3]W^6$$

- $k = 3$:

$$X_3 = x[0] + x[1]W^3 + x[2]W^6 + x[3]W^9$$

This set of four equations can be expressed more conveniently, in matrix form, as:

$$\begin{bmatrix} X_0 \\ X_1 \\ X_2 \\ X_3 \end{bmatrix} = \begin{bmatrix} 1 & 1 & 1 & 1 \\ 1 & W^1 & W^2 & W^3 \\ 1 & W^2 & W^4 & W^6 \\ 1 & W^3 & W^6 & W^9 \end{bmatrix} \begin{bmatrix} x[0] \\ x[1] \\ x[2] \\ x[3] \end{bmatrix} \quad (5.9)$$

Expanding W^1, W^2, W^3 etc. using Euler's identity, or by considering the periodicity* of these terms, you should be able to show that:

---

*Periodicity: Remember that $e^{j\theta}$ represents a unit vector on the Argand diagram, inclined at an angle of θ to the positive real axis. It follows that $e^{\frac{-j2\pi}{4}}$ and $e^{\frac{-j18\pi}{4}}$, i.e. W^1 and W^9, only differ in that W^9 has made two extra revolutions. Effectively, they both represent unit vectors inclined at $-90°$, or $+270°$, to the positive real axis, and so they are equal to $-j$.)

$W^1 = W^9 = -j$, $W^2 = W^6 = -1$, $W^3 = j$ and $W^4 = 1$

Substituting into equation 5.9:

$$\begin{bmatrix} X_0 \\ X_1 \\ X_2 \\ X_3 \end{bmatrix} = \begin{bmatrix} 1 & 1 & 1 & 1 \\ 1 & -j & -1 & j \\ 1 & -1 & 1 & -1 \\ 1 & j & -1 & -j \end{bmatrix} \begin{bmatrix} x[0] \\ x[1] \\ x[2] \\ x[3] \end{bmatrix}$$

Expanding the matrix equation:

$$X_0 = x[0] + x[1] + x[2] + x[3] \tag{5.10a}$$

$$X_1 = x[0] - jx[1] - x[2] + jx[3] \tag{5.10b}$$

$$X_2 = x[0] - x[1] + x[2] - x[3] \tag{5.10c}$$

$$X_3 = x[0] + jx[1] - x[2] - jx[3] \tag{5.10d}$$

If we now let:

$$A(0) = x[0] + x[2]$$

$$A(1) = x[0] - x[2]$$

$$B(0) = x[1] + x[3]$$

$$B(1) = x[1] - x[3]$$

then:

$$X_0 = A(0) + B(0) \tag{5.11a}$$

$$X_1 = A(1) - jB(1) \tag{5.11b}$$

$$X_2 = A(0) - B(0) \tag{5.11c}$$

$$X_3 = A(1) + jB(1) \tag{5.11d}$$

When carried out by a computer, the process of multiplication is much more time consuming than the process of addition and so it is important to reduce the number of multiplications in the algorithm as much as possible. We have clearly done this and so the computation of X_0 to X_3 should be achieved much more quickly than if the DFT were to be used.

The FFT, represented by the four equations (5.11a–d) above, can be depicted by the signal flow diagram of Fig. 5.20. These FFT signal flow diagrams are often referred to as 'butterfly' diagrams for obvious reasons.

We have looked at the FFT for just four samples but the FFT exists for *any* number of samples (although the algorithm is simplest for an even number). Indeed, the relative improvement in processing time increases as N increases. It can be shown that the number of multiplications performed by the normal DFT, when transforming a signal consisting of N samples, is of the order of N^2, whereas, for the FFT, it is roughly $N \log_2 N$. For our four-sample sequence this corresponds to approximately 16 multiplications for the DFT and eight for the FFT. However,

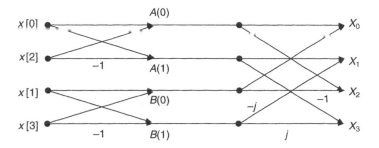

Figure 5.20

if we take a more realistic number of samples, such as 1024, then the DFT carries out just over one million multiplications and the FFT just over 10 000, i.e. a much more significant difference. See Proakis and Manolakis (1996), Lynn and Fuerst (1994), or Ingle and Proakis (1997) if you require a more detailed treatment of this topic.

N.B.   We are spending time looking at the FFT and its inverse, the IFFT, because these transforms are used in the design of digital signal processing systems. However, before moving on from the FFT it should be pointed out that digital signal processors, in turn, are used to *find* the FFT of a signal and so allow its spectrum to be analysed. This is a very common application of DSP systems.

The fast inverse Fourier transform

So far we have concentrated on the FFT but the fast inverse Fourier transform is equally important. As it is the inverse process to the FFT, basically, all that is needed is to move through the corresponding FFT signal flow diagram *in the opposite direction*. However, it can be shown that we first need to change the signs of the exponentials in the W terms – in other words, change the sign of the phase angle of the multipliers. As phase angles of $\pm 180°$, $\pm 540°$, $\pm 900°$ etc. are effectively the same, as are $\pm 0°$, $\pm 360°$, $\pm 720°$ etc. then all that is needed is to change the sign of the j multipliers. This is because unit vectors with phase angles of $\pm 90°$ or $\pm 270°$ etc., are *not* the same, one is represented on an Argand diagram by j and the other by $-j$.

It can also be shown that the resulting outputs need to be divided by N.

Figure 5.21 shows the modified butterfly diagram for our four-sample system. Just to show that this works, consider $x[0]$. From Fig. 5.21:

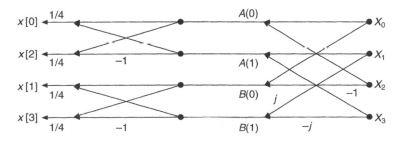

Figure 5.21

$x[0] = (A(0) + A(1))/4$

but $A(0) = X_2 + X_0$ and $A(1) = X_1 + X_3$

$\therefore x[0] = (X_2 + X_0 + X_1 + X_3)/4$

Substituting for X_0, X_1, X_2 and X_3 using equations 5.10 (a–d), you should find that you end up with $x[0] = x[0]$, which shows that the algorithm represented by Fig. 5.21 works! (See Damper (1995) and also Ifeachor and Jervis (1993) for further details.)

5.16   MATLAB AND THE FFT

The FFT for a sequence can be obtained *very* easily by using the MATLAB, 'fft' function. The following program will generate the sampled, complex frequency values for the 10 sample sequence 1, 2, 3, 5, 4, 2, –1, 0, 1, 1.

```
x=[1 2 3 5 4 2 -1 0 1 1];    % sampled signal values
y=fft(x)                      % generates corresponding
                              % sampled frequency values
```

The sampled magnitude and phase values are displayed graphically in Fig. 5.22, using a sampling frequency of 10 kHz.

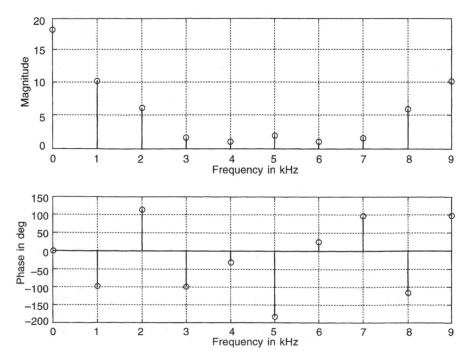

Figure 5.22

More interestingly, Fig. 5.23(a) shows the sampled signal produced by adding three sine waves of unit amplitude, having frequencies of 100, 210 and 340 Hz, while Fig. 5.23(b) displays the sampled amplitude spectrum, obtained using the 'fft' function. The sampling frequency used was 1 kHz and 50 samples were taken and so the frequency resolution is 20 Hz, (f_s/N). Note that the signals of frequencies 100 and 340 Hz show up very clearly in Fig. 5.23(b), while the 210 Hz signal is not so obvious. This is because 100 and 340 Hz are two of the sampled frequencies, i.e. nf_s/N, while 210 Hz is not. Even so, the fft makes a valiant attempt to reveal it!

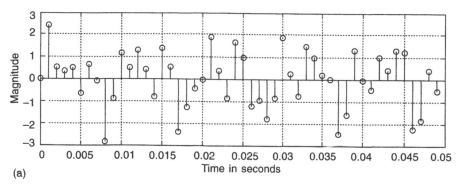

(a)

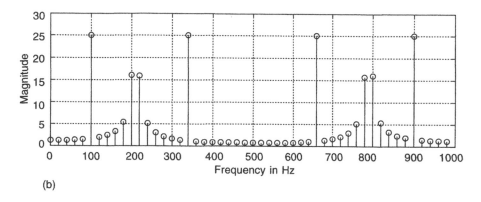

(b)

Figure 5.23

The program is:

```
T=0.001;                        % sampling frequency of
                                % 1 kHz
f1=100; f2=210; f3=340;         % frequencies of the
                                % three signals
n=(0:49);                       % 50 samples taken
t=n*T;                          % times at which signal
                                % sampled
y=sin(2*pi*f1*t) + sin
(2*pi*f2*t) + sin(2*pi*f3*t); % y is sum of the three
                                % signals
```

```
subplot(2,1,1);
stem(t,y)                       % plot y against t using
                                % "stems"
grid                            % add grid
xlabel('Time in seconds')
fs=n*20                         % sampled f values (freq
                                % resolution = 20 Hz)
subplot(2,1,2)
m=abs(fft(y));                  % find magnitude values
                                % only, i.e not phase
stem(fs,m)                      % plot m against frequency
grid
xlabel ('Frequency in Hz'),ylabel('Magnitude')
```

There is also a MATLAB 'ifft' function. The following simple program shows how the 10-point sequence, 1, 2, 3, 5, etc., 'fft'd earlier, can be recovered by using the ifft function.

```
x=[1 2 3 5 4 2 -1 0 1 1]; % sampled sequence
y=fft(x);                  % find fft
z=ifft(y)                  % find ifft of the sampled
                           % spectral content
```

The results obtained are:

'1+0i, 2+0i, 3+0i, 5+0i . . .'

i.e. the original real, sampled signal of 1, 2, 3, 5, etc. is revealed by the ifft function.

5.17   RECAP

•   The DFT and IDFT are extremely useful but have the disadvantage that they require a relatively large amount of computation.

•   The fast Fourier transform (FFT), and its inverse, have been developed to reduce the processing time.

5.18   A FINAL WORD OF WARNING

In this book we have been looking at the theory behind the analysis and design of digital filters. However, when you tackle some *practical* processing you might find that the system doesn't perform quite as expected! The serious problem of pole and zero 'bunching' was mentioned in the previous chapter, but there are many other sources of error which could affect the behaviour of digital signal processing systems. One obvious cause of inaccuracies is the quantization of the input signal due to the A-to-D converter. The digital output values of the converter will only be approximate representations of the continuous input values – the smaller the number of A-to-D bits the more this quantization error. Further errors are introduced because the processor registers have a finite number of bits in which to store the various data, such as input, output and multiplier values. This will result in numbers being truncated or rounded. These errors will be catastrophic if, as a result of an addition or a multiplication for example, the resulting values are just too big to be stored completely. Such an overflow of data usually results in the most significant bits being lost and so the remaining stored data are nonsense. To reduce the risk of this, the processing software often scales data, i.e. reduces them. Under certain circumstances, the rounding of data values in IIR systems can also cause oscillations, known as 'limit cycles'.

This is just a very brief overview of some of the more obvious shortcomings of DSP systems. If you would like to learn more about this aspect of digital signal processors then some good sources are Jackson (1996), Proakis and Manolakis (1996), Terrell and Lik-Kwan Shark (1996) and Ifeachor and Jervis (1993).

5.19 CHAPTER SUMMARY

In this, the final chapter, a case for using FIR rather than IIR filters was made. The major advantage of FIR systems is that it is possible to design them to have linear phase characteristics, i.e the phase shift of any frequency component, caused by passing through the filter, is proportional to the frequency. These linear-phase filters effectively delay the various signal frequency components by the same time, and so avoid signal distortion. For an FIR filter to have a linear phase response, it must be designed to have symmetric transfer function coefficients.

Apart from it being impossible for an FIR filter to be unstable (unlike IIR filters), a further advantage is that they can be designed to have a magnitude response of any shape – although only 'brickwall' profiles have been considered here.

Two major design techniques were described – the 'Fourier' or 'windowing' method and also the 'frequency sampling' approach. The improvement in filter performance, resulting from the use of a suitable 'window' profile, such as Hamming, Hanning or Blackman, was also discussed and demonstrated.

As the two design methods rely heavily on either the continuous or the discrete Fourier transforms – or their inverses, these topics were also covered. The problem of the relatively long processing time required by the DFT and IDFT was then considered and the alternative, much faster FFT and FIFT described.

Finally, some of the more obvious sources of filter error were highlighted.

5.20 PROBLEMS

1 A lowpass FIR filter is to be designed using the 'Fourier' or 'windowing' method. The filter should have 11 coefficients and a cut-off frequency of

$0.25f_N$. The filter is to be designed using (a) a rectangular window and (b) a Hamming window.

2 A *bandpass* FIR filter, of length nine (nine coefficients) is required. It is to have lower and upper cut-off frequencies of 3 and 6 kHz respectively and is intended to be used with a sampling frequency of 20 kHz. Find the filter coefficients using the Fourier method with a rectangular window.

3 A discrete signal consists of the finite sequence 1, 0, 2. Use the DFT to find the frequency components within the signal, assuming a sampling frequency of 1 kHz. Check that the transformation has been carried out correctly by applying the IDFT to the derived frequency components in order to recover the original signal samples.

4 Use the 'frequency sampling' method to design a lowpass FIR filter with a cut-off frequency of 500 Hz, given that the filter will be used with a sampling frequency of 4 kHz. The frequency sampling is to be carried out at eight points. (N.B. An *even* number of sampling points.)

5 A *bandstop* filter is to be designed using the frequency sampling technique. The lower and upper cut-off frequencies are to be 1 and 3 kHz respectively and the filter is to be used with a sampling frequency is 9 kHz. The frequency sampling is to be carried out at nine points.

6 A discrete signal consists of the four samples 1, 2, 0, –1.

   (a) Use the FFT butterfly diagram of Fig. 5.20 to find the sampled frequency spectrum for the signal.

   (b) Apply the signal flow diagram of Fig. 5.21, i.e. the fast inverse Fourier transform, to the frequency samples obtained in (a) above, to reveal the original sampled signal.

Answers to self-assessment tests and problems

Chapter 1

Section 1.9

As this is a running average filter which averages the current and previous *two* input samples, then:

$$\text{Data out} = \frac{\text{Current input sample plus the previous two inputs}}{3}$$

Applying this equation, and remembering that the two storage registers are initially reset, the corresponding input and output values will be as shown in Table S1.1. The first two output values of 0.67 and 1.67 are not strictly valid as the two zeros, initially stored in the two storage registers, are still feeding through the system.

Table S1.1

Time	Input data	Output data
0	2	0.67
T	3	1.67
2T	6	3.67
3T	2	3.67
4T	1	3.00
5T	7	3.33
6T	4	4.00
7T	1	4.00
8T	2	2.33
9T	4	2.33

The input and output values are also shown in Fig. S1.1. Notice how the averaging tends to smooth

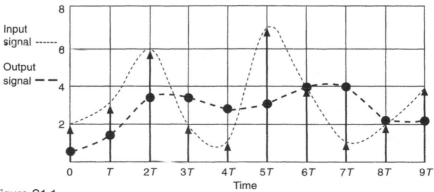

Figure S1.1

out the variations of the input signal. This is a good demonstration that the running average filter is a type of lowpass filter.

Section 1.11

(a) From Fig. 1.11:

Data out = Data in $-0.1A - 0.4B + 0.2C$

(b) Table S1.2 shows all outputs, and also the A, B and C values, corresponding to the eight input samples.

Table S1.2

Time	Data in	A	B	C	Data out
0	3.00	0.00	0.00	0.00	3.00
T	5.00	3.00	3.00	0.00	3.50
2T	6.00	5.00	3.50	3.00	4.70
3T	6.00	6.00	4.70	3.50	4.22
4T	8.00	6.00	4.22	4.70	6.65
5T	10.00	8.00	6.65	4.22	7.38
6T	0.00	10.00	7.38	6.65	-2.62
7T	0.00	0.00	-2.62	7.38	2.53

Problems

1. (a) Not programmable.

   (b) Unlikely that two 'identical' analogue processors will behave in an identical way.

   (c) Affected by ageing and temperature changes.

   (d) Not as versatile.

2. True.

3. (a).

4. Recursive, feedback.

5. False (true for IIR filters – running average is an FIR filter).

6. False (only true for IIR filters).

7. (a) 40 000 bytes; (b) 5 kHz (Nyquist frequency = half sampling frequency).

8. 3, 4.4, 6.9, 6.1, 3.9, 3.2.

9. 2.00, 2.60, 4.56, 3.18, −0.47, 0.19.

10. 1, 1, 1.2, 1.4, 1.64, 1.92.

   It looks suspiciously as though the output will continue to increase with time, i.e. that the system is unstable. (In fact, this particular system *is* unstable and needs redesigning.)

Chapter 2

Section 2.4

1. (a) $x(t) = 2 \cos (3\pi t)$

   $\therefore \ \omega = 3\pi = 2\pi f$

$\therefore f = 1.5\,\text{Hz}$

To avoid aliasing, this signal must be sampled at a minimum of 3 Hz, and so 20 Hz is fine.

(b) As the sampling frequency is 20 Hz, the samples are the signal values at 0, 0.05, 0.1, 0.15, 0.2 and 0.25 s.

$\therefore x[n] = 2, 1.78, 1.17, 0.31, -0.62, -1.42$

(c) $x[n - 4] = 0, 0, 0, 0, 2, 1.78, 1.17, 0.31, -0.62, -1.42.$

2. (a) $x[n] + w[n] = 1, 0, 4, -3$

(b) $3w[n - 2] = 3(0, 0, -1, -2, 1, -4) = 0, 0, -3, -6, 3, -12$

$\therefore x[n] + 3w[n - 2] = 2, 2, 0, -5, 3, -12$

(c) $0.5x[n - 1] = 0, 1, 1, 1.5, 0.5$

$\therefore x[n] - 0.5x[n - 1] = 2, 1, 2, -0.5, -0.5$

3. (a) From Fig. 2.8, the sequence at A is $-2(x[n - 1]) = 0, -4, -2, -6, 2.$

(b) $y[n] = x[n] - 2(x[n - 1]) = 2, -3, 1, -7, 2.$

Section 2.8

1. (a) $X(z) = x[0] + x[1]z^{-1} + x[2]z^{-2} + \ldots = 1 + 2z^{-1} + 4z^{-2} + 6z^{-4}.$

(b) $x[n] = \dfrac{n}{(n + 1)^2}$

$\therefore x[0] = 0, x[1] = 1/4$ etc.

$\therefore X(z) = \dfrac{1}{4}z^{-1} + \dfrac{2}{9}z^{-2} + \dfrac{3}{16}z^{-3} + \ldots$

(c) $z^{-2}X(z) = z^{-2} + 2z^{-3} + 4z^{-4} + 6z^{-6}$

(d) $x[n] = e^{-0.2n}$

$\therefore x[0] = 1, x[1] = 0.82$ etc.

$\therefore X(z) = 1 + 0.82z^{-1} + 0.67z^{-2} + 0.55z^{-3}$

2. (a) $\dfrac{3z}{z - e^{-2T}} = \dfrac{3z}{z - 0.819}$

(b) $\dfrac{2z \sin(0.3)}{z^2 - 2z \cos(0.3) + 1}$

(c) $\dfrac{0.1ze^{-0.2}}{(z - e^{-0.2})^2} = \dfrac{0.0819z}{(z - 0.819)^2}$

(d) $\dfrac{0.1z}{(z - 1)^2}$

Section 2.10

1. $X(z) = 1 - z^{-1}$

$Y(z) = 1 + 2z^{-1} - z^{-2} - 2z^{-3}$

$T(z) = \dfrac{Y(z)}{X(z)} = \dfrac{1 + 2z^{-1} - z^{-2} - 2z^{-3}}{1 - z^{-1}}$

Carrying out the polynomial division should give:

$T(z) = 1 + 3z^{-1} + 2z^{-2}$

2. (a) $Y(z) = (1 + 3z^{-1})(1 + 2z^{-2}) = 1 + 3z^{-1} + 2z^{-2} + 6z^{-3}$
Therefore, the output sequence is 1, 3, 2, 6.

(b) $Y(z) = (1 + 3z^{-1})(2 + 3z^{-1} - z^{-2})$
giving an output sequence of 2, 9, 8, –3.

3. (a) $Y(z) = \dfrac{z}{z - 0.6}$

From the z-transform tables:

$$0.6 = e^{-aT}$$

but $y(t) = e^{-at}$

$\therefore y[n] = e^{-anT}$ (sampled sequence)

$\therefore y[n] = (0.6)^n$

$\therefore Y(z) = 1 + 0.6z^{-1} + 0.36z^{-2} + \ldots$

(b) $Y(z) = \dfrac{z}{(z - 0.6)}\dfrac{z}{(z - 1)} = \dfrac{z^2}{(z - 0.6)(z - 1)}$

$\therefore \dfrac{Y(z)}{z} = \dfrac{z}{(z - 0.6)(z - 1)} = \dfrac{A}{z - 0.6} + \dfrac{B}{z - 1}$

This gives $A = -1.5$ and $B = 2.5$

$\therefore \dfrac{Y(z)}{z} = -\dfrac{1.5}{z - 0.6} + \dfrac{2.5}{z - 1}$

$\therefore Y(z) = 2.5\dfrac{z}{z - 1} - 1.5\dfrac{z}{z - 0.6}$

$\therefore y[n] = 2.5u[n] - 1.5(0.6)^n$

$\therefore y[n] = 1 + 1.6z^{-1} + 1.96z^{-2} + \ldots$

Section 2.16

1. $Y(z) = X(z) - 0.8X(z)z^{-1} + 0.5Y(z)z^{-1}$
$\therefore Y(z)(1 - 0.5z^{-1}) = X(z)(1 - 0.8z^{-1})$

$\therefore \dfrac{Y(z)}{X(z)} = T(z) = \dfrac{1 - 0.8z^{-1}}{1 - 0.5z^{-1}}$

Dividing $(1 - 0.8z^{-1})$ by $(1 - 0.5z^{-1})$ you should get $1 - 0.3z^{-1} - 0.15z^{-2} \ldots$ and so the first three terms of the output sequence are 1, –0.3 and –0.15.

2. From Fig. S2.1:

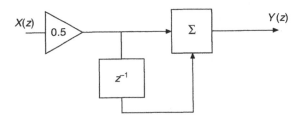

Figure S2.1

$Y(z) = 0.5X(z) + 0.5X(z)z^{-1}$
$\therefore T(z) = 0.5(1 + z^{-1})$

As the z-transform for a unit ramp is $0 + z^{-1} + 2z^{-2} + 3z^{-3} + \ldots$, then the z-transform of the system output is given by:

$Y(z) = 0.5(1 + z^{-1})(z^{-1} + 2z^{-2} + \ldots) = 0 + 0.5z^{-1} + 1.5z^{-2} + \ldots$

3. $Y(z) = 0.2(X(z) + X(z)z^{-1} + X(z)z^{-2} + X(z)z^{-3} + X(z)z^{-4})$ (S2.1)

$\therefore Y(z)z^{-1} = 0.2(X(z)z^{-1} + X(z)z^{-2} + X(z)z^{-3} + X(z)z^{-4} + X(z)z^{-5})$ (S2.2)

Equation (S2.2) – equation (S2.1):

$Y(z)z^{-1} - Y(z) = 0.2X(z)z^{-5} - 0.2X(z)$
$Y(z)(1 - z^{-1}) = 0.2X(z)(1 - z^{-5})$

$\therefore \dfrac{Y(z)}{X(z)} = T(z) = 0.2\dfrac{1 - z^{-5}}{1 - z^{-1}}$

i.e. this is the transfer function for a recursive filter.

4. $V(z) = X(z) + b_1V(z)z^{-1} + b_2V(z)z^{-2}$

$\therefore V(z)(1 - b_1z^{-1} - b_2z^{-2}) = X(z)$

$\therefore V(z) = \dfrac{X(z)}{1 - b_1z^{-1} - b_2z^{-2}}$ (S2.3)

Also, $Y(z) = V(z) + a_1V(z)z^{-1} + a_2V(z)z^{-2}$

$= V(z)(1 + a_1z^{-1} + a_2z^{-2})$ (S2.4)

From equations (S2.3) and (S2.4):

$T(z) = \dfrac{1 + a_1z^{-1} + a_2z^{-2}}{1 - b_1z^{-1} - b_2z^{-2}}$

Problems

1. 1, 3, 0, –2.

2. $T(z) = 1 + z^{-1} + 2z^{-2}$. Unit step response: 1, 2, 4, 4.

3. $T(z) = (1 + 0.5z^{-1})/(1 - 0.5z^{-1}) = 1 + z^{-1} + 0.5z^{-2} + 0.25z^{-3}$.

Unit sample response: 1, 1, 0.5, 0.25.

4. Output sequence: 1, 0, 1, 0, –2. [$T(z) = (1 - z^{-2})$].

5. $T(z) = 0.5(1 - 0.4z^{-1})/(1 - z^{-2})$. (Hint: The input to the ×0.5 attenuator is $2Y(z)$.)

6. 0, 0.8, 0.96, 0.992

$[(1 + z^{-1} + z^{-2} + z^{-3} + \dots) - (1 + 0.2z^{-1} + 0.04z^{-2} + 0.008z^{-3} + \dots)]$

Chapter 3

Section 3.4

1. $Y(s) = \dfrac{2(s + 2)}{s(s^2 + 4s + 3)}$

There is therefore a single zero at $s = -2$. There is a pole at $s = 0$ and also at $s = -3$ and -1 (roots of $s^2 + 4s + 3$). The pole–zero diagram is shown in Fig. S3.1. The pole at A will correspond to a step of height k_1, while the poles at B and C represent exponentially decaying d.c. signals of k_2e^{-t} and k_3e^{-3t} respectively, where k_1, k_2 and k_3 are multipliers. (Although it *is* possible to find the k-values from the p–z diagram, this is not expected here.) Figure S3.2 shows the MATLAB plots for the signals corresponding to the three pole positions (k-values of 1 have been used for all three signals, as it is only general signal shapes that are required).

$\dfrac{2(s + 2)}{s(s + 1)(s + 3)} = \dfrac{a}{s} + \dfrac{b}{s + 1} + \dfrac{c}{s + 3}$

which gives $a = 4/3$, $b = -1$ and $c = -1/3$. Therefore, from the Laplace transform tables:

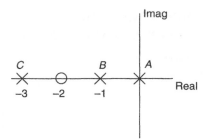

Figure S3.1

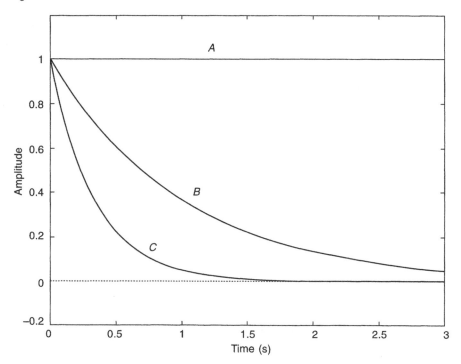

Figure S3.2

$$y(t) = \frac{4}{3} - e^{-t} - \frac{1}{3} e^{-3t}$$

2. This transform has no zeros but has poles at $s = \pm j2$ and so this represents a non-decaying sine wave of angular frequency 2. From the Laplace transform tables:

$$y(t) = 3 \sin 2t$$

i.e. the transform first needs to be rearranged to $3 \times 2/(s^2 + 2^2)$. It can then be recognized as a standard transform.

Section 3.11

1. This p–z diagram represents an exponentially decaying d.c. signal – the signal samples falling to 0.8 of the previous sample – the pulse train being delayed by one sampling period. Carrying out the polynomial division, i.e. dividing z^{-1} by $1 - 0.8z^{-1}$, results in the series $z^{-1} + 0.8z^{-2} + 0.64z^{-3} + \ldots$ (N.B. This assumes no 'pure' gain in the system, i.e. that $T(z) = k/(z - 0.8)$ *where* $k = 1$. It is impossible to know the true value of k from the p–z diagram alone.)

2. The Nyquist frequency (f_N) = 60 Hz, i.e. $180° \equiv 60$ Hz.

$$\therefore f_A = \frac{150}{180} \times 60 = 50 \text{ Hz}, f_B = f_N = 60 \text{ Hz}, f_C = 30 \text{ Hz}, f_D = 0 \text{ Hz}$$

3. $z = e^{sT}$

Pole A:

$s = \sigma + j\omega$
$\therefore s = -1$
$\therefore z_A = e^{-T} = e^{-0.5} = 0.61$

Pole-pair B1 and B2:

$z = e^{-(1 \pm j)0.5} = e^{-0.5}e^{\pm j0.5} = 0.61e^{\pm j0.5}$

This represents conjugate poles 0.61 from the origin, the pole vectors to the origin being inclined at angles of ±0.5 radians from the positive real axis.

Pole C:

$s = 0$
$\therefore z = 1$

Pole-pair D1 and D2:

$z = e^{\pm j0.5}$

i.e. the pole vectors have a magnitude of 1 and are inclined at ±0.5 radians (±28.6°) to the positive real axis.

Pole E:

$z = e^{0.5} = 1.65$

The z-plane p–z diagram is shown in Fig. S3.3.

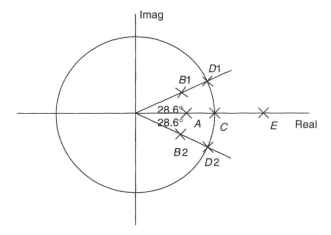

Figure S3.3

Section 3.15

$$T(s) = \frac{4(s+1)(s+2)}{(s^2 + 2s + 2)}$$

Therefore there are real s-plane zeros at $s = -1$ and -2 and complex conjugate poles at $-1 \pm j$ (roots of the denominator). Figure S3.4 shows the p–z diagram.

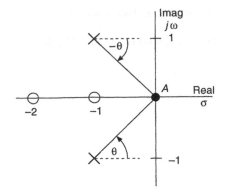

Figure S3.4

0 rad/s (Point A):

$$\text{Magnitude} = \frac{k \prod \text{zero distances}}{\prod \text{pole distances}}$$

$$= \frac{4 \times 2}{\sqrt{2}\sqrt{2}} = 4 \text{ (12 dB)}$$

Phase angle = Σ zero angles $- \Sigma$ pole angles = 0

1 rad/s (Point B, Fig. S3.5):

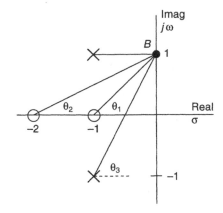

Figure S3.5

$$\text{Magnitude} = \frac{4 \times \sqrt{2}\sqrt{5}}{1 \times \sqrt{5}} = 5.66$$

Phase angle = $\theta_1 + \theta_2 - \theta_3 = \tan^{-1} 1 + \tan^{-1} 0.5 - \tan^{-1} 2 = 8.2°$

A similar approach should show that the gain and phase angle at 2 rad/s are 5.66 (15 dB) and −8.2°.

For very large frequencies the distances from the frequency point to the two poles and the two zeros are almost identical and so the gain must be very close to 4. Further, as the angles made by the zero and pole vectors approach 90° then the phase angle must be zero. Figure S3.6 shows the Bode plot for the system.

Section 3.18

$$T(z) = \frac{z}{(z^2 + z + 0.5)(z - 0.8)}$$

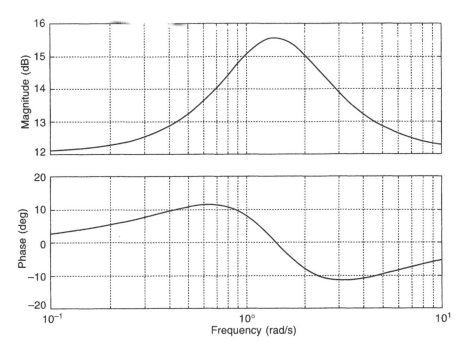

Figure S3.6

One zero at $z = 0$ and three poles at $z = 0.8$ and $-0.5 \pm j0.5$.

The p–z diagram is shown in Fig. S3.7.

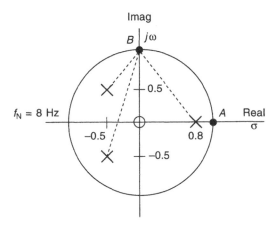

Figure S3.7

As the Nyquist frequency is 8 Hz, points A and B must represent signal frequencies of 0 and 4 Hz respectively.

(a) $f = 0$ Hz:

By using

$$\text{Magnitude} = \frac{k \, \Pi \text{ zero distances}}{\Pi \text{ pole distances}}$$

and

Phase angle = Σ zero angles – Σ pole angles

it should be fairly easy to show that the magnitude and phase angle of the frequency response at 0 Hz are 2 and 0° respectively.

$f = 4$ Hz:

From Fig. S3.7:

$$\text{Magnitude} = \frac{1}{\sqrt{0.25 + 0.25}\sqrt{0.25 + 2.25}\sqrt{1 + 0.64}} = 0.7$$

Phase angle = $90 - (\tan^{-1}1 + \tan^{-1}3 + \tan^{-1}(-1/0.8)) = -155.3°$ or $204.7°$

(b) $f = 0$ Hz:

Substituting $z = 1$ into $T(z) = \dfrac{z}{(z^2 + z + 0.5)(z - 0.8)}$:

$$T(z) = \frac{1}{(2.5)(0.2)} = 2\angle 0$$

$f = 4$ Hz:

Substituting $z = j$ into $T(z) = \dfrac{z}{(z^2 + z + 0.5)(z - 0.8)}$:

$$T(z) = \frac{j}{(-1 + j + 0.5)(j - 0.8)}$$

$$= \frac{j}{(-0.5 + j)(-0.8 + j)}$$

$$= \frac{j}{-0.6 - 1.3j}$$

$$= \frac{1\angle 90°}{1.43\angle 245.2°}$$

$$= 0.7\angle - 155.2°$$

Problems

1. (a) Zero at 0.1 and pole at 0.3.

   (b) Gain at 0 Hz = 1.29, phase angle = 0.
   Gain at 250 Hz = 0.96, phase angle = $-11°$ (or 349°).

   (c) Gain at 375 Hz = 0.8, phase angle = $-6.1°$ (353.9°).

2. (a) $T(z) = \dfrac{(z - 0.819)}{(z - 0.905)(z - 0.741)}$.

   (b) The d.c. gains of the continuous and digital filters are approximately 0.67 and 7.36 respectively. It follows that the transfer function of the digital filter would have to be scaled by multiplying by 0.67/7.36.

3. (a) As the z-transform for the unit sample function is 1, then

   $$Y(z) = T(z) = \frac{1}{z^2 - z + 0.5}$$

   i.e. no zeros but poles at $0.5 \pm j0.5$.

   (b) The z-transform of a unit step is $z/(z - 1)$. It follows that

   $$Y(z) = \frac{1}{(z^2 - z + 0.5)} \frac{z}{(z - 1)}$$

i.e. p–z diagram as for (a) above, but an extra pole at +1 and a zero at the origin. Gain at 0 Hz = 2 and at $f_s/4 = 0.89$.

4. Dividing z^2 by $z + 1$ (or z by $1 + z^{-1}$) results in the z-transform of $z - 1 + z^{-1} - z^{-2} + \ldots$ for the unit sample response. The initial z indicates that the output sequence begins one sampling period before the input arrives! Impossible!

5. $p = 0.79$ ($p^2 = 0.62$).
The gain at 0 Hz is approximately 1.2.
The filter is a bandstop filter, with maximum gains of 1.2 at 0 and 100 Hz (f_N) and minimum gain of 0.5 (–6 dB) at 50 Hz.

Chapter 4

Section 4.5

$$T(z) = \frac{k}{z^2 + a^2}$$

There must therefore be two poles at $z = \pm ja$. Figure S4.1 shows the two poles and also points A, B and C corresponding to signal frequencies of 100, 90 and 110 Hz respectively. (N.B. The positions of points B and C on the unit circle are not drawn to scale.)

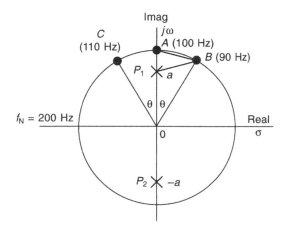

Figure S4.1

Using

$$\text{Gain} = \frac{k \, \Pi \text{ zero distances}}{\Pi \text{ pole distances}}$$

At 100 Hz (Point A):

$$1 = \frac{k}{(1 - a)(1 + a)} \qquad (S4.1)$$

At 90 Hz (or 110 Hz):

The gain = –3 dB

$$\therefore 0.71 \approx \frac{k}{P_1 B(1 + a)}$$

('$(1 + a)$' as B is very close to A)

To find P_1B:

Arc $AB = 1 \times \theta = \dfrac{10}{200}\pi = 0.16$

As the angle $\angle AOB$ is relatively small ($\approx 9°$), arc $AB \approx$ line AB. Also, triangle P_1AB approximates to a right-angled triangle, with $\angle P_1AB \approx 90°$.

$\therefore P_1B \approx \sqrt{(1-a)^2 + 0.16^2}$

$\therefore 0.71 \approx \dfrac{k}{(1+a)\sqrt{(1-a)^2 + 0.16^2}}$ \hfill (S4.2)

Dividing equation (S4.2) by (S4.1), and carrying out a few lines of algebra, you should eventually find that $a \approx 0.86$. Substituting this value back into equation (S4.1) or (S4.2) should give $k \approx 0.26$.

$\therefore T(z) = \dfrac{0.26}{z^2 + 0.86^2} = \dfrac{0.26}{z^2 + 0.74}$

Section 4.9

$T(s) = \dfrac{16}{(s+4)^2}$

First check whether pre-warping is necessary:

$\omega_c' = \dfrac{2}{T}\tan\left(\dfrac{\omega T}{2}\right)$

$\therefore \omega_c' = \dfrac{2}{0.2}\tan\left(\dfrac{4 \times 0.2}{2}\right) = 4.23$

This is approximately a 5% difference compared to ω_c, and so pre-warping *is* necessary.

$T(s) = \dfrac{\dfrac{16}{16}}{\left(\dfrac{s}{4}+1\right)^2} = \dfrac{1}{\left(\dfrac{s}{4}+1\right)^2}$

Replacing all s/ω_c terms with s/ω_c':

$T(s)' = \dfrac{1}{\left(\dfrac{s}{4.23}+1\right)^2} = \dfrac{17.89}{(s+4.23)^2}$

Applying the bilinear transform:

$T(z) = \dfrac{17.89}{\left(10\left(\dfrac{z-1}{z+1}\right)+4.23\right)^2}$

After a few lines of algebra:

$T(z) = 0.088\,\dfrac{(z+1)^2}{z^2 - 0.81z + 0.164}$

Section 4.11

(a) $T(s) = \dfrac{4}{(s+2)^2}$

From the *s/z* transform tables:

$$T(z) = \frac{0.4ze^{-0.2}}{(z - e^{-0.2})^2} = \frac{0.328z}{z^2 - 1.64z + 0.672}$$

Scaling the *z*-domain transfer function so that the d.c. gains of the two filters become the same:

$$\left| \frac{4}{(s+2)^2} \right|_{s=0} = k \left| \frac{0.328z}{z^2 - 1.64z + 0.672} \right|_{z=1}$$

$$\therefore 1 = \frac{0.328k}{1 - 1.64 + 0.672}$$

$\therefore k \approx 0.098$

$$\therefore T(z) = \frac{0.032z}{z^2 - 1.64z + 0.672}$$

(b) $T(s) = \dfrac{2}{(s+1)(s+2)}$

Applying the method of partial fractions:

$$\frac{2}{(s+1)(s+2)} = \frac{A}{s+1} + \frac{B}{s+2}$$

Using the 'cover-up' rule (or any other method):

$A = 2, B = -2$

$$\therefore T(s) = \frac{2}{s+1} - \frac{2}{s+2}$$

From the *s/z* transform tables:

$$T(z) = \frac{2z}{z - e^{-0.1}} - \frac{2z}{z - e^{-0.2}} = \frac{2z}{z - 0.9} - \frac{2z}{z - 0.82}$$

After a few lines of algebra, you should find that

$$T(z) = \frac{0.172z}{z^2 - 1.724z + 0.741}$$

and, after equalizing the d.c. gains of the two filters:

$$T(z) = \frac{0.0172z}{z^2 - 1.724z + 0.741}$$

Section 4.13

(a) $T(s) = \dfrac{4}{(s+2)^2}$

This particular continuous filter has no zeros and a double pole at $s = -2$.
From $z = e^{sT}$, the equivalent pole positions in the *z*-plane must be given by $z = e^{-0.2} = 0.82$.

$$\therefore T(z) = \frac{k(z+1)^2}{(z - 0.82)^2}$$

(A double zero has been added at $z = -1$ to balance up the number of poles and zeros.)

After equalizing the d.c. gains of the filters, the final transfer function is given by:

$$T(z) = \frac{0.0081(z+1)^2}{(z-0.82)^2}$$

(b) $T(s) = \dfrac{s+2}{(s+1)(s+3)}$

i.e. a zero at $s = -2$ and two poles at $s = -1$ and -3. These transform to a zero in the z-plane at $z = 0.819$ and poles at 0.905 and 0.741.

$$\therefore T(z) = \frac{k(z-0.819)}{(z-0.905)\,(z-0.741)}$$

or

$$T(z) = \frac{0.091(z-0.819)}{(z-0.905)(z-0.741)}$$

after scaling.

Section 4.18

$$T(z) = 0.15\,\frac{z+1}{z-0.7}$$

$\omega_c = 2\pi f_c = 110\pi,\ \Omega_c = 200\pi,\ T = 10^{-3}$ s

$$a = -\frac{\cos\left[(\omega_c - \Omega_c)T/2\right]}{\cos\left[(\omega_c + \Omega_c)T/2\right]} = -\frac{\cos\left[(110\pi - 200\pi)\times 0.5\times 10^{-3}\right]}{\cos\left[(110\pi + 200\pi)\times 0.5\times 10^{-3}\right]} = -1.12$$

$$z \rightarrow -\frac{1+az}{z+a} = -\frac{1-1.12z}{z-1.12} = \frac{1.12z-1}{z-1.12}$$

$$\therefore T(z) = 0.15\,\frac{\dfrac{1.12z-1}{z-1.12}+1}{\dfrac{1.12z-1}{z-1.12}-0.7}$$

After some tidying up:

$$T(z) = 0.757\left(\frac{z-1}{z-0.514}\right)$$

Problems

1. $T(z) = 0.71\left(\dfrac{z-1}{z-0.42}\right).$

2. (a) $T(z) = 0.853\left(\dfrac{z-0.77}{z-0.51}\right).$

    (b) $T(z) = \dfrac{0.04(z+1)^2}{(z-0.512)(z-0.675)}$

3. $T(z) = 0.652\left(\dfrac{z-1}{z-0.368}\right).$

4. (a) $T(z) = \dfrac{0.15z}{(z-0.45)^2}.$

    (b) $T(z) = \dfrac{0.544z(z-0.575)}{(z-0.67)(z-0.3)}.$

5. (a) $T(s)_{\text{Highpass}} = \dfrac{s}{s+6}$, $T(z)_{\text{Highpass}} = 0.761\left(\dfrac{z-1}{z-0.55}\right)$.

(b) $T(z)_{\text{Lowpass}} = 0.13\left(\dfrac{z+1}{z-0.741}\right)$, $T(z)_{\text{Highpass}} = 0.758\left(\dfrac{z-1}{z-0.514}\right)$.

Chapter 5

Section 5.9

1. (a) $f_c = 3$ kHz, $f_N = 6$ kHz

From Section 5.6:

$$C_n = \frac{\sin\left(n\pi\dfrac{f_c}{f_N}\right)}{n\pi} = \frac{\sin(0.5n\pi)}{n\pi} \tag{S5.1}$$

Substituting $n = -5$ to $+5$ into equation (S5.1) gives coefficients of:

0.063662, 0, –0.106103, 0, 0.31831, 0.5, 0.31831, 0, –0.106103, 0, 0.063662

$$\therefore T(z) = 0.063662 - 0.106103z^{-2} + 0.31831z^{-4} + 0.5z^{-5} + 0.31831z^{-6}$$
$$- 0.106103z^{-8} + 0.063662z^{-10}$$

(b) The Hanning window function is given by:

$$w(n) = \frac{1}{2}\left[1 - \cos\left(\frac{2n\pi}{N-1}\right)\right] \tag{S5.2}$$

Substituting $n = 0$ to 10 into equation (S5.2) gives Hanning coefficients of :

0, 0.095492, 0.345492, 0.654508, 0.904508, 1, 0.904508, 0.654508, 0.345492, 0.095492, 0

Multiplying the filter coefficients of part (a) by the corresponding Hanning coefficient gives a modified filter transfer function of:

$$T(z) = -0.036658z^{-2} + 0.287914z^{-4} + 0.5z^{-5} + 0.287914z^{-6} - 0.036658z^{-8}$$

or, more sensibly:

$$T(z) = -0.036658 + 0.287914z^{-2} + 0.5z^{-3} + 0.287914z^{-4} - 0.036658z^{-6}$$

2. The filter coefficients are the same as those for the lowpass filter designed in Section 5.6, before applying the Hamming window function, apart from (a) the signs being changed and (b) the central coefficient being $1 - (f_c/f_N) = 0.6$ (see Section 5.8).

Section 5.13

The required magnitude response and the sampled frequencies are shown in Fig. S5.1. From Section 5.12, equation (5.8b), the filter coefficients, $x[n]$, are given by:

$$x[n] = \frac{1}{N}\left[\sum_{k=1}^{(N-1)/2} 2|X_k| \cos\left[2\pi k(n-\alpha)/N\right] + X_0\right] \tag{S5.3}$$

where N is the number of frequency samples taken (nine here), $\alpha = (N-1)/2$, i.e. 4, and $|X_k|$ is the magnitude of the kth frequency sample. In this example $|X_0|$, $|X_1|$ and $|X_2| = 0$, while $|X_3| = 0.5$ and $|X_4| = 1$. Substituting into equation (S5.3)

$$x[0] = \frac{1}{9}[2(0.5 \cos(6\pi(0-4)/9) + \cos(8\pi(0-4)/9))] = -0.016967$$

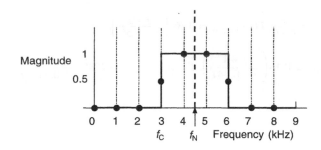

Figure S5.1

$$x[1] = \frac{1}{9}[2(0.5\cos(6\pi(1-4)/9) + \cos(8\pi(1-4)/9))] \approx 0$$

Similarly, $x[2] = 0.114677$, $x[3] = -0.264376$ and $x[4] = 0.333333$

$$\therefore T(z) = -0.016967 + 0.114677z^{-2} - 0.264376z^{-3} + 0.333333z^{-4} - 0.264376z^{-5}$$
$$+ 0.114677z^{-6} - 0.016967z^{-8}$$

Problems

1 (a) Filter coefficients: −0.0450, 0, 0.0750, 0.1592, 0.2251, 0.25,
                0.2251, 0.1592, 0.0750, −0.0450.

   (b) Filter coefficients: −0.0036, 0, 0.0298, 0.1086, 0.2053, 0.25,
                0.2053, 0.1086, 0.0298, 0, −0.0036.

2. Filter coefficients: 0.1225, −0.0952, −0.2449, 0.0452, 0.3,
                0.0452, −0.2449, −0.0952, 0.1225.

3. Magnitude and phase angle of the three frequency components of 0, 500 Hz and 1500 Hz are $3\angle 0°$, $1.732\angle 90°$ and $1.732\angle -90°$ respectively.

4. Filter coefficients: −0.0176, −0.0591, 0.1323, 0.4444, 0.4444,
                0.1323, −0.0591, −0.0176.

5. Filter coefficients: −0.0103, 0.0556, 0.2451, −0.0681, 0.5556,
                −0.0681, 0.2451, 0.0556, −0.0103.

6. The FFT values are: 2, $1 - j3$, 0, $1 + j3$, i.e. the magnitudes of the four frequency components, 0, $f_s/4$, $f_s/2$ and $3f_s/4$ are 2, $\sqrt{10}$, 0 and $\sqrt{10}$ respectively, while the corresponding phase angles are 0, $\tan^{-1}(-3)$, 0, $\tan^{-1}3$.

References and Bibliography

Bolton, W. 1998: *Control engineering.* Harlow: Addison-Wesley Longman.

Carlson, G.E. 1998: *Signal and linear system analysis.* Chichester: John Wiley.

Chassaing, R., Horning, D.W. 1990: *Digital signal processing with the TMS320C25.* Chichester: John Wiley.

Damper, R.I. 1995: *Introduction to discrete-time signals and systems.* London: Chapman & Hall.

Denbigh, P. 1998: *System analysis and signal processing.* Harlow: Addison-Wesley.

Dorf, R.C., Bishop, R.H. 1995: *Modern control systems.* Harlow: Addison-Wesley.

El-Sharkawy, M. 1996: *Digital signal processing applications with Motorola's DSP56002 processor.* London: Prentice-Hall.

Etter, D.M. 1997: *Engineering problem solving with MATLAB.* London: Prentice-Hall.

Hanselman, D., Littlefield, B. 1997: *The student edition of MATLAB – version 5.* London: Prentice-Hall.

Hayes, M.H. 1999: *Digital signal processing.* London: McGraw-Hill.

Howatson, A.M. 1996: *Electrical circuits and systems – an introduction for engineers and physical scientists.* Oxford: Oxford University Press.

Ifeachor, E.C., Jervis, B.W. 1993: *Digital signal processing – a practical approach.* Harlow: Addison-Wesley.

Ingle, V.K., Proakis, J.G. 1991: *Digital signal processing laboratory using the ADSP-2101 microcomputer.* London: Prentice-Hall.

Ingle, V.K., Proakis, J.G. 1997: *Digital signal processing using MATLAB v.4.* London: PWS.

Jackson, B.J. 1996: *Digital filters and signal processing.* London: Kluwer Academic.

Johnson, J.R. 1989: *Introduction to digital signal processing.* London: Prentice-Hall.

Jones, D.L., Parks, T.W. 1988: *A digital signal processing laboratory using the TMS32010.* London: Prentice-Hall.

Ludeman, L.C. 1987: *Fundamentals of digital signal processing.* Chichester: John Wiley.

Lynn, P.A., Fuerst, W. 1994: *Introductory digital signal processing with computer applications.* Chichester; John Wiley.

Marven, C., Ewers, G. 1994: *A simple approach to digital signal processing.* Texas Instruments.

Meade, M.L., Dillon, C.R. 1991: *Signals and systems – models and behaviour.* London: Chapman & Hall.

Millman, J., Grabel, A. 1987: *Microelectronics.* London: McGraw-Hill.

Oppenheim, A.V., Schafer, R.W. 1975: *Digital signal processing.* London: Prentice-Hall.

Powell, R. 1995: *Introduction to electric circuits.* London: Arnold.

Proakis, J.G., Manolakis, D.G. 1996: *Digital signal processing – principles, algorithms and applications.* London: Prentice-Hall.

Steiglitz, K. 1995: *A digital signal processing primer – with applications to digital audio and computer music*. Harlow: Addison-Wesley.

Terrell, T.J. 1980: *Introduction to digital filters*. London: Macmillan.

Terrell, T.J., Lik-Kwan, S. 1996: *Digital signal processing – a student guide*. London: Macmillan.

The MathWorks Inc. 1995: *The student edition of MATLAB*. London: Prentice-Hall.

Virk, G.S. 1991: *Digital computer control system*. London: Macmillan.

Appendix A Some useful Laplace and z-transforms

Time function		Laplace transform	z-transform
$\delta(t)$	(unit impulse)	1	1
$\delta(t - nT)$	(delayed unit impulse)	e^{-nsT}	z^{-n}
$u(t)$	(unit step)	$\dfrac{1}{s}$	$\dfrac{z}{z-1}$
t	(unit ramp)	$\dfrac{1}{s^2}$	$\dfrac{Tz}{(z-1)^2}$
t^2		$\dfrac{2}{s^3}$	$\dfrac{T^2 z(z+1)}{(z-1)^3}$
e^{-at}		$\dfrac{1}{s+a}$	$\dfrac{z}{z - e^{-aT}}$
$1 - e^{-aT}$		$\dfrac{a}{s(s+a)}$	$\dfrac{z(1 - e^{-aT})}{(z-1)(z - e^{-aT})}$
te^{-at}		$\dfrac{1}{(s+a)^2}$	$\dfrac{Tze^{-aT}}{(z - e^{-aT})^2}$
$t - \dfrac{1}{a}(1 - e^{-at})$		$\dfrac{a}{s^2(s+a)}$	$\dfrac{Tz}{(z-1)^2} - \dfrac{(1 - e^{-aT})z}{a(z-1)(z - e^{-aT})}$
te^{-at}		$\dfrac{1}{(s+a)^2}$	$\dfrac{Tze^{-aT}}{(z - e^{-aT})^2}$
$\sin \omega t$		$\dfrac{\omega}{s^2 + \omega^2}$	$\dfrac{z \sin \omega T}{z^2 - 2z \cos \omega T + 1}$
$\cos \omega t$		$\dfrac{s}{s^2 + \omega^2}$	$\dfrac{z(z - \cos \omega T)}{z^2 - 2z \cos \omega T + 1}$
$e^{-aT} \sin \omega T$		$\dfrac{\omega}{(s+a)^2 + \omega^2}$	$\dfrac{ze^{-aT} \sin \omega T}{z^2 - 2ze^{-aT} \cos \omega T + e^{-2aT}}$
$e^{-aT} \cos \omega T$		$\dfrac{s+a}{(s+a)^2 + \omega^2}$	$\dfrac{z(z - e^{-aT} \cos \omega T)}{z^2 - 2ze^{-aT} \cos \omega T + e^{-2aT}}$

Appendix B Frequency transformations in the *s*- and *z*-domains

s-DOMAIN TRANSFORMATIONS

To transform from a lowpass filter with cut-off frequency ω:

Required filter type	Replace *s* with;
Lowpass	$\dfrac{\omega s}{\Omega_c}$
Highpass	$\dfrac{\omega \Omega_c}{s}$
Bandstop	$\dfrac{s\omega(\Omega_h - \Omega_1)}{s^2 + \Omega_1\Omega_h}$
Bandpass	$\dfrac{\omega(s^2 + \Omega_1\Omega_h)}{s(\Omega_h - \Omega_1)}$

where Ω_c is the new cut-off frequency for the low and highpass filters, and Ω_1 and Ω_h are the lower and higher cut-off frequencies for the bandstop and bandpass filters.

z-DOMAIN TRANSFORMATIONS

To transform from a lowpass filter with cut-off frequency ω:

Required filter type	Replace z with:	where:
Lowpass	$\dfrac{1-az}{z-a}$	$a = \dfrac{\sin(\omega - \Omega_c)T/2}{\sin(\omega + \Omega_c)T/2}$
Highpass	$-\dfrac{1+az}{z+a}$	$a = -\dfrac{\cos(\omega - \Omega_c)T/2}{\cos(\omega + \Omega_c)T/2}$
Bandstop	$\dfrac{1 - \dfrac{2az}{1+b} + z^2\dfrac{1-b}{1+b}}{\dfrac{1-b}{1+b} - \dfrac{2az}{1+b} + z^2}$	$a = \dfrac{\cos(\Omega_h + \Omega_l)T/2}{\cos(\Omega_h - \Omega_l)T/2}$ $b = \tan\left[(\Omega_h - \Omega_l)T/2\right]\tan\left(\dfrac{\omega T}{2}\right)$
Bandpass	$-\dfrac{1 - \dfrac{2abz}{1+b} + z^2\left(\dfrac{b-1}{1+b}\right)}{\dfrac{b-1}{1+b} - \dfrac{2abz}{1+b} + z^2}$	a – as for the bandstop filter $b = \cot\left[(\Omega_h - \Omega_l)T/2\right]\tan\left(\dfrac{\omega T}{2}\right)$

where Ω_c is the new cut-off frequency for the low and highpass filters, and Ω_l and Ω_h are the lower and higher cut-off frequencies for the bandstop and bandpass filters.

Index

9 780750 650489